The Na$^+$, K$^+$ pumps keep us going

The Na$^+$, K$^+$ pumps keep us going

How passive and active transport of sodium (Na$^+$) and potassium (K$^+$) control performance and fatigue in skeletal muscle.

By Torben Clausen,
professor emeritus, MD,
Department of Biomedicine,
Aarhus University

Aarhus University Press

The Na⁺, K⁺ pumps keep us going
© Torben Clausen and Aarhus University Press 2016
Cover: Nethe Ellinge Nielsen, Trefold
Layout and typesetting: Narayana Press
Publishing Editor: Simon Olling Rebsdorf
This book is typeset in Candida & Grotesque
and printed on Arctic Volume White 130 g
Printed by Narayana Press, Denmark

Printed in Denmark 2016

ISBN 978 87 7124 939 2

Aarhus University Press
Langelandsgade 177
DK-8200 Aarhus N
Denmark
www.unipress.dk

Published with the financial support of Aarhus
University Research Foundation

International distributors:
Gazelle Book Services Ltd.   ISD
White Cross Mills   70 Enterprise Drive, Suite 2
Hightown, Lancaster, LA1 4XS   Bristol, CT 06010
United Kingdom   USA
www.gazellebookservices.co.uk   www.isdistribution.com

**PEER
REVIEWED**

/ In accordance with requirements of the Danish Ministry of Higher Education and Science, the certification
means that a PhD level peer has made a written assessment justifying this book's scientific quality.

Contents

Preface

This book approaches readers with an interest in biology and sports and who have perhaps been planning to do research or to study physiology and biochemistry. With figures, diagrams, curves and tables, the book seeks to explain how a single molecule, the Na^+, K^+ pump, creates the conditions required for muscles to work. This happens by the coordination and control of the work of the billions (10^{12}) of Na^+, K^+ pumps transporting Na^+ and K^+ across the membranes of muscle cells. This vital process continues every second, day and night. It requires energy, but in return allows the muscles to develop their often considerable work in a goal-directed and dirigible manner. This control is exerted by the nervous system acting on the skeletal muscle cells, the contraction of which we are responsible for. In addition, the activity and number of Na^+, K^+ pumps in the muscles are regulated by hormones, by training or inactivity, and are failing in connection to a number of diseases. Thus, the Na^+, K^+-pumps are favouring the working capacity of the muscles, performance, and fitness. Knowledge about their function promotes the understanding of muscle physiology and function, causes of fatigue, endurance, and capacity for maximum performance. In short, the Na^+, K^+ pumps keep us going.

The author was trained as an MD and has 50 years of experience with teaching and research in muscle and general physiology as assistant professor and full professor of human physiology at the health faculty of Aarhus University in Denmark. Professor Jens Christian Skou was supervising his doctoral thesis. He has published 215 articles about physiology, which have been quoted around 10.000 times in the international literature. The information presented in this book is based upon a selection of these publications, their illustrations and those of other authors working in the same area.

01

The Na$^+$, K$^+$ pump and its discovery

The Na$^+$, K$^+$ pump was discovered in 1957 by professor, MD Jens Christian Skou at the Institute of Physiology, Aarhus University. He detected the pumps in the nerves of crabs from the Danish Sea, and later they were identified and confirmed in many other tissues of animals and human subjects (see the first comprehensive review by Skou, 1965). The discovery of the Na$^+$, K$^+$-pump and its explanatory power was a breakthrough, also in general membrane transport physiology. Therefore, in 1997, it was awarded by the Nobel Prize in Chemistry. Meanwhile, it has given rise to thousands of papers, books, lectures and meetings around the world, the discovery of other transport systems and clarification of the pathophysiology of numerous clinical disorders.

◀ Jens Christian Skou on his bike in Aarhus University Park, October 1997.

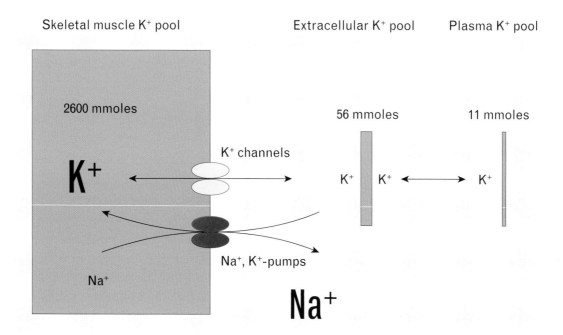

Fig. 1. The extrarenal distribution of K⁺ in the human body. By far the largest pool of K⁺ is confined to the skeletal muscle cells, from where it may rapidly be released into the blood plasma and the extracellular space. (Clausen, 2015).

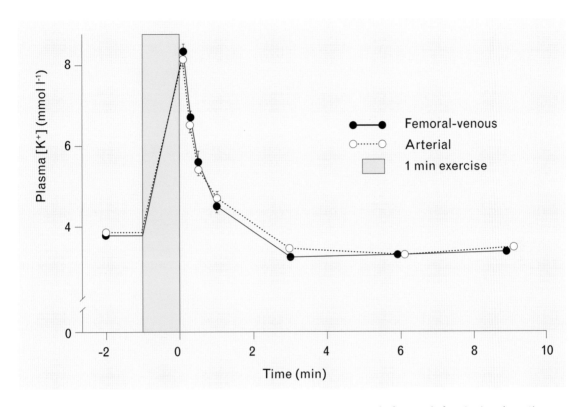

Fig. 2. Arterial and femoral-venous plasma potassium concentration before and after 1 min exhausting exercise (means ± S.E.M of measurements on 12 subjects (Medbø and Sejersted, *J. Physiol.* 421, 105-122, 1990).

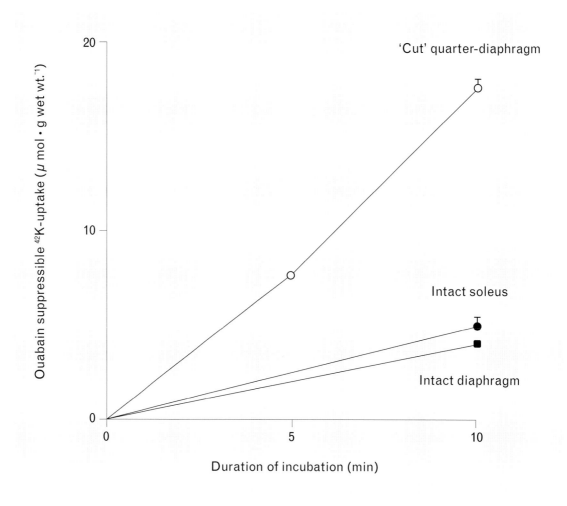

Fig. 3. Effect of tissue integrity on ouabain-supressible ^{42}K uptake in mouse muscle. Quarterdiaphragm, intact soleus and diaphragm muscles were washed and preincubated for 15 min in Krebs-Ringer bicarbonate buffer without and with ouabain (10^{-3} M). They were then incubated for 5 or 10 min in the same buffer containing ^{42}K, blotted and counted for ^{42}K. Results are given as difference between amount of ^{42}K taken up in the absence and presence of ouabain, with bars indicating SE. (Biron et al., *J. Physiol.* Vol. 297, p. 52. 1979).

▶▶ **Fig. 4.** Na$^+$, K$^+$ transport pathways in skeletal muscle. The transmembrane Na$^+$, K$^+$ concentration gradients are generated by the continuous operation of the Na$^+$, K$^+$-pumps. The Na$^+$ gradient drives the sym-port of amino acids and chloride and the anti-ports of H$^+$ and Ca^{2+}. By far the major influx pathways for Na$^+$ are the voltage-sensitive Na$^+$ channels. Passive K$^+$ fluxes are mediated by the voltage-sensitive inward rectifier (inw.) and delayed rectifier (del.), the Ca^{2+} sensitive and the ATP-sensitive K$^+$ channels. (Modified from Clausen, *Physiol. Reviews*, 2003).

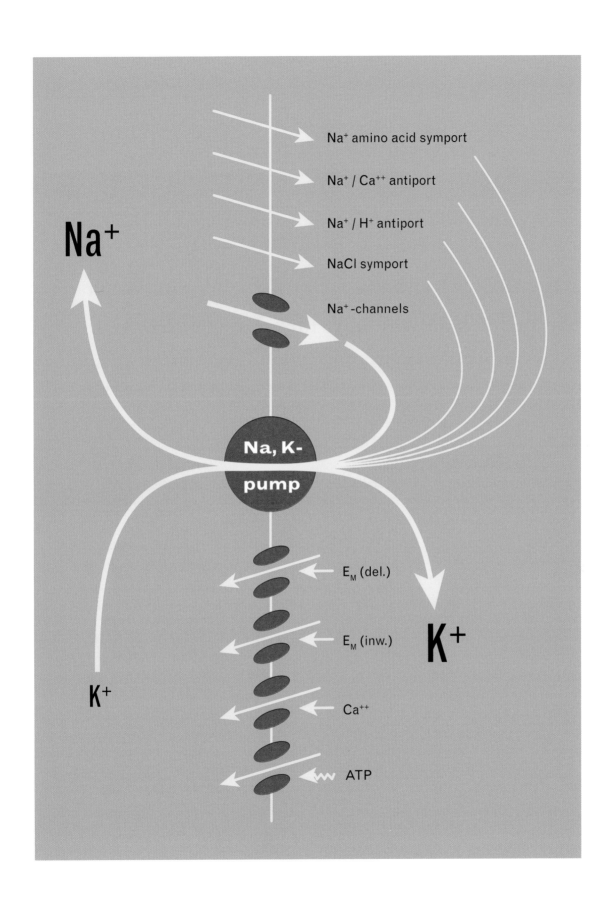

02

The Transport and distribution of Na$^+$ and K$^+$ in skeletal muscle and how they are quantified

As shown in Fig. 1 (the front page of the book) one of the most striking contrasts in the body is the distribution of K$^+$, where the intracellular pool of skeletal muscles contains 2.600 mmoles of K$^+$, amounting to 46-fold the K$^+$ content of the extracellular pool and 236-fold the K$^+$ content of the blood plasma. This implies that even modest losses of K$^+$ from the muscle cells may cause relatively large increases in the concentration of K$^+$ in these rather narrow extracellular spaces. Thus, as demonstrated by Medbø & Sejersted (1990), and as described in detail in an innovative comprehensive review (Sejersted & Sjøgaard, 2000), illustrated in Fig. 2, only 1 min of intense exercise is sufficient to double the concentration of K$^+$ in the plasma of femoral venous as well as arterial blood. This means that the heart is directly exposed to K$^+$ at a concentration sufficient to cause cardiac arrest if it were under resting conditions. Furthermore, the post-excitatory hyperkalemia inhibits excitability of the muscle cells, limiting further work and thus causes fatigue. Fortunately, flooding of the extracel-

lular phase with K$^+$ is counterbalanced by the Na$^+$, K$^+$-pumps, which are working every second, day and night. This book seeks to explain how this single molecule creates the conditions required for muscles to work. As expressed in the title of one of our papers, the Na$^+$, K$^+$ pumps protect muscle excitability and contractility during exercise (Nielsen & Clausen, 2000). Knowledge about their function promotes the understanding of muscle physiology and function, causes of fatigue, endurance, and capacity for maximum contractile performance.

It should be noted, however, that following the large exercise-induced rise in plasma K$^+$ the Na$^+$, K$^+$-pump-mediated clearance of K$^+$ is rapid enough to cause hypokalemia (see Fig. 2), which represents an even larger risk of cardiac arrest. A prospective study on 21,481 healthy male physicians showed 23 sudden deaths from cardiac causes associated with vigorous exercise (Albert et al. 2000). An electrocardiographic study on 21 squash players demonstrated that 2/3rds showed arrhythmia during or immediately after the play (Northcote et al., 1986). The relative risk of sudden death was seven-fold higher among those who rarely engaged in vigorous exercise (Albert et al., 2000).

Based on measurements of ^{3}H-ouabain binding sites in human skeletal muscle (300 pmoles/g wet wt. = 0.3 nmole/g wet wt = 0.3 µmole/kg wet wt.), the content of Na^+, K^+-pumps amounts to 0.3µmol/kg muscle wet wt. With a maximum rate of ATP-splitting per mole of Na^+, K^+-pumps of 8.000 ATP molecules/min may take place and with a number of 2 K^+ ions being transported per cycle, the maximum capacity of the Na^+, K^+-pumps in skeletal muscles for accumulation of K^+ amounts to 0.3x28x8.000x2 µmoles/min = 134.4 mmoles K^+/min, more than twice the amount of K^+ in the extracellular pool.

As shown in Fig 1, the extracellular pool of K^+ in human muscle is 56 mmoles. This amount of K^+ may be cleared in 56/134 = 0.417 min = 25 sec. Even though this is not often likely to happen, these values give an idea of the maximum capacity of the Na^+, K^+-pumps.

Measurements of Na^+, K^+ contents and the rates of Na^+, K^+ transport in skeletal muscle is primarily performed using isolated intact muscles and muscle cells, but may also be done in vivo. Over the last century, various techniques have been developed, first using isolated frog and toad muscles incubated in buffer. In the forties the small and thin cut rat and mouse hemi-diaphragm muscles were introduced with the motive of improving the access of oxygen to the cells. However, since most of the cut muscle cells in these preparations were leaky, the passive fluxes of Na^+ and K^+ were much in excess of those in intact diaphragm muscles, causing considerable overestimation (4.4-fold) of the ouabain-suppressible ^{42}K uptake (see Fig. 3 in

Biron et al., 1979). Therefore, in experiments with cut hemidiaphragm preparations, the thermogenic action of thyroid hormones is likely to be overestimated. Intact muscle fibers utilize only 4-10 % of their total energy turnover for active Na^+, K^+ transport (Biron et al.; 1979; Clausen et al., 1991). In more recent studies small intact animal muscles, e. g. soleus, extensor digitorum longus (EDL), flexor digitorum brevis, epitrochlearis and sternohyoideus muscles are used (Mu et al. 2011). Quite recently, bundles of muscle fibers were isolated from the abdominal rectus muscle in patients undergoing abdominal surgery. These preparations weighed around 600 mg and when incubated in Krebs-Ringer buffer at 30 °C turned out to maintain force (when tested with 2 s stimulation at 60 Hz) for at least 12 h in vitro (Skov et al., 2015). The standard conditions for incubation of isolated muscles are in Krebs-Ringer bicarbonate buffer (KR) at room temperature or 30-37 °C during continuous gassing with a mixture of 5 % CO_2 and 95 % of O_2, allowing the maintenance of a constant pH of 7.4.

The incubation may also take place at 30 °C in air without buffer allowing the maintenance of force and Na^+, K^+ contents comparable to those in KR buffer (Clausen, 2008a+b). The isotopic tracers ^{22}Na, ^{24}Na, ^{42}K, ^{86}Rb and flame photometry were used to quantify the fluxes of Na^+ and K^+ in µmoles per g wet wt. The intracellular accumulation of the isotopes may be followed over time intervals from seconds to hours if followed by a 4x15 min washout in ice-cold buffer allowing washout of extracellular isotopes with correction for loss of intracellular isotopes (Clausen & Kohn, 1977; Buchanan et al., 2002).

As shown in Fig. 4, the transmembrane Na^+, K^+ concentration gradients are generated by the continuous operation of the Na^+, K^+ pumps, moving Na^+ ions actively out of the cell (from right to left) and K^+ ions actively into the cell (from left to right). By far the major passive influx pathways for Na^+ are the voltage sensitive Na^+ channels. The concentration gradient for Na^+ drives the sym-port of amino acids into the cell, the anti-port of Ca^{2+} out of the cell (from right to left), the anti-port of H^+ ions out of the cell and the symport of $NaCl$ into the cell (see Skou's scheme of ion fluxes across the cell membranes in Fig. 39). Passive K^+ fluxes are mediated by the voltage-sensitive inward rectifier (inw.), the delayed rectifier (del.), the Ca^{2+}-sensitive and the ATP-sensitive K^+ channels. The Na^+, K^+ pump mediated uptake of K^+ (using ^{42}K or ^{86}Rb) or efflux of Na^+ (using ^{22}Na) may be quantifed by pre-exposure to ouabain for 15 min, usually at a concentration (10^{-3} M), sufficient to induce complete inhibition of the Na^+, K^+ pumps. The Na^+, K^+ pump-mediated uptake of K^+ and efflux of Na^+ are quantified by deducting the value obtained in the presence of ouabain from that obtained in untreated muscles (Clausen, 2010).

The abovementioned Na^+ channels may be blocked selectively by saxitoxin or tetrodotoxin at micromolar concentrations. The high affinity of these agents for the Na^+ channels allow them (when labelled with ^3H) to be used for the quantification of the Na^+ channels in nerves or muscles (Harrison et al., 1997). Such measurements are performed by incubating intact skeletal muscles or strips of muscles weighing 20-40 mg at 4 °C for 120 min in buffer containing $2.5x10^{-8}$ M [^3H]saxitoxin (0.05µCi/ml) (sufficient to give complete saturation of the specific saxitoxin binding sites. (Harrison et al. 1997; Harrison & Clausen, 1998).

^{42}K and ^{86}Rb are both used as isotopic tracers for K^+, and as shown in Fig. 5, the Na^+, K^+ pump-mediated (ouabain-suppressible) uptake of these two tracers for K^+ are closely similar, both when stimulated by insulin, salbutamol, CGRP and when inhibited by graded concentrations of ouabain (Dørup & Clausen, 1994). Because the half-life of ^{42}K is very short (12.5 hrs), ^{86}Rb (half-life 18.7 days) is often preferred. However, the efflux (fractional loss) of ^{86}Rb from preloaded rat soleus is 2.3 times smaller than that of ^{42}K. Salbutamol and CGRP increase ^{86}Rb efflux, but inhibit ^{42}K efflux. Finally, the inhibitory effect of bumetanide on ^{86}Rb uptake gives rise to the false impression that skeletal muscle contains an active $NaKCl_2$ co-transport system. These discrepancies imply that for studies of K^+ efflux and bumetanide-sensitive K^+ transport, ^{86}Rb is not even an acceptable tracer for the detection of changes K^+ fluxes in skeletal muscle (Dørup & Clausen, 1994).

In conclusion, control experiments with ^{42}K are essential for any accurate characterization of unknown K^+ transport processes or their response to hormones or pharmaceuticals. Likewise, isotope fluxes and contents (of ^{22}Na, ^{24}Na, ^{42}K and ^{86}Rb) should be compared with flame photometric measurements of the contents of Na^+ and K^+.

03

Rapid passive fluxes of Na$^+$ and K$^+$ start muscle contraction and set a limit to excitability

The action potentials that start the contraction of a skeletal muscle are initiated by a rapid influx of Na$^+$ ions taking place within fractions of seconds, immediately followed by an almost equivalent efflux of K$^+$ ions (Hodgkin & Horowicz, 1959; Clausen et al., 2004). This generates a rapid sequence of depolarization and repolarization, described as an action potential, the signal that starts the release of Ca^{2+} ions from the sarcoplasmic reticulum into the cytoplasm (Fig. 6). This, in turn, leads to the activation of the contractile filaments which are instrumental in generating the muscle contraction.

The cost of the rapidity of this exchange of Na$^+$ and K$^+$ is that within a few seconds, the concentration of K$^+$ in the narrow extracellular spaces may rise sufficiently to cause spreading depolarization of the muscle cells, leading to a rapid loss of excitability and contractility. K$^+$ is cleared from the extracellular space by diffusion into the capillaries, which by an action of K$^+$ on the capillary walls (Fig. 7) are simultaneously relaxed and dilated (DeClerck et al., 2003) increasing

blood flow up to considerably above resting value (see also Dua et al., 2009). Besides, the elevation of [K$^+$]$_o$ may trigger localized release of the peptide hormone calcitonin gene related peptide (CGRP) (Sakaguchi et al., 1991), which is one of the most potent vasodilatatory agents, allowing improved transfer of K$^+$ into the blood stream.

Another efficient mechanism for the clearance of K$^+$ from the extracellular space are the Na$^+$, K$^+$ pumps, which are markedly activated during muscle contraction, allowing up to 20-fold increase in transport rate within 10 seconds with utilization of all available Na$^+$, K$^+$ pumps in the muscle (Fig. 8) (Nielsen & Clausen, 1997). This constitutes the major transport challenge for the Na$^+$, K$^+$ pumps, and due to the 3/2 coupling of Na$^+$ efflux and K$^+$ influx, the Na$^+$, K$^+$ pumps contribute to the restoration of the membrane potential and excitability of the muscle cells. During intense exercise, however, the loss of K$^+$ from the working muscles may exceed the transport capacity of the Na$^+$, K$^+$ pumps, leading to an increase in the extracellular concentration of K^+ and depolarization, which inhibits further excitation and contractile activity. As shown in Fig 7 and 9, this sequence of events is an important cause of muscle fatigue. Thus, during continuous work, there is a close correlation between the rise in extracellular K^+ and the rate of force decline (Fig. 9) (Clausen et al., 2004; Clausen, 2015). This explains why the Na^+, K^+ pumps are essential for the maintenance of working capacity. Acute inhibition of the Na^+, K^+ pumps with cardiac glycosides or down-regulation of the Na^+, K^+-pump capacity by K^+-deficiency may severely impair contractile performance. Intoxication with cardiac glycosides in patients gives rise to hyperkalemia and fatigue which is the predominant symptom of this disorder (Bismuth et al., 1973). Likewise, K^+ deficiency is a common cause of muscle fatigue that can be related to the concomitant down-regulation of the content of Na^+, K^+ pumps in skeletal muscles (Nørgaard et al., 1981; Clausen, 1998; Clausen, 2003). This is of considerable general interest, because K^+ deficiency is one of the most common nutritional disorders in the 3. world. Besides, millions of patients all over the world use diuretics every day for the treatment of hypertension, cardiac or lung diseases, which often cause K^+-deficiency with down-regulation of the Na^+, K^+-pumps in their muscles (Dørup et al., 1988; Dørup, 1994). Alone in Denmark, 350.000 patients are treated with diuretics (Dørup 1994) and around half of those have developed K^+-deficiency and reduced content of Na^+,K^+-pumps in skeletal muscles. Several other disorders (diabetes, hypothyroidism, immobilization, heart failure, starvation, McArdle disease and muscular dystrophy) decrease the content of Na^+, K^+-pumps in skeletal muscles which can often be associated with muscle fatigue (Clausen, 1998).

Conversely, numerous studies show that training leads to up-regulation of the content of Na^+, K^+-pumps in skeletal muscles (Kjeldsen et al., 1986c; Clausen, 2003b). This is associated with more efficient clearance of K^+ from blood plasma during exercise (McKenna et al., 1993; Bangsbo et al., 2009). The content of Na^+, K^+ pumps in skeletal muscle is upregulated by thyroid hormones (Kjeldsen et al., 1986a; Everts & Clausen, 1988; Clausen, 2003b; Riis et al., 2005) and the glucocorticoid dexamethasone (Dørup & Clausen, 1997; Nordsborg et al., 2005; Hostrup et al., 2016). Whereas thyroid hormones cause fatigue, dexamethasone may increase endurance (Nordsborg et al., 2008).

Measurement of the rates of Na^+, K^+ transport in skeletal muscle is primarily performed using isolated intact muscles and muscle cells, but may also be done in vivo. Over the last century, various techniques have been developed, first using isolated frog and toad muscles or frog semitendinous muscle single fibres incubated in buffer. In the 1940ties the small and thin cut rat and mouse hemi-diaphragm muscles were introduced with the motive of improving the access of oxygen to the cells. However, since most of the cut muscle cells in these preparations were leaky, the passive fluxes of Na^+ and K^+ were much in excess of those in intact diaphragm muscles, causing considerable overestimation (4.4-fold) of the ouabain-suppressible ^{42}K uptake (Fig. 3). Therefore, in experiments with cut hemi-diaphragm preparations, the thermogenic action of thyroid hormones is likely to be overestimated. Intact muscle fibers utilize only 4-10 % of their total energy turnover for active Na^+, K^+ transport (Biron et al., 1979; Clausen T, Van Hardeveld C, and Everts ME, 1991).

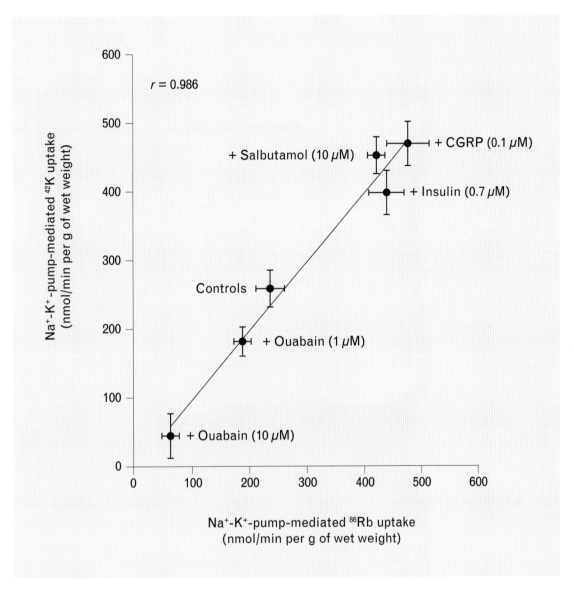

Fig. 5. Correlation between Na[+], K[+] pump-mediated ^{86}Rb and [42]K uptake in rat soleus muscle.
After equilibration for 30-60 min, soleus muscles were preincubated in Krebs-Ringer bicarbonate buffer without or with ouabain (10^{-3} M). Then they were incubated in buffer containing [42]K (0.2 µCi/ml and [86]Rb (0.2µCi/ml) without or with ouabain and the indicated additions for 10 min. Two groups of muscles were pre-exposed for 15 min to ouabain at concentrations (10^{-6}-10^{-5} M) producing partial inhibition of the Na[+], K[+]-pumps. After the 10 min incubation with the isotopes, the muscles were blotted, weighed and homogenized in 2 ml of 0.3 M trichloroacetic acid for counting. The [42]K and ^{86}Rb activity of the supernatant obtained by centrifugation of the homogenate was first determined by measurement of Cerenkov radiation. After decay of [42]K, ^{86}Rb activity was determined and [42]K activity could be calculated by subtraction. Na[+], K[+] -pump mediated ^{86}Rb uptake was calculated as the difference between uptake measured in the absence and the presence of 10^{-3} M ouabain. Each point represents the difference ± SEM between means of measurements performed on four to eight muscles incubated without or with 10^{-3} M ouabain. (Clausen et al., *J. Physiol.*, 1987).

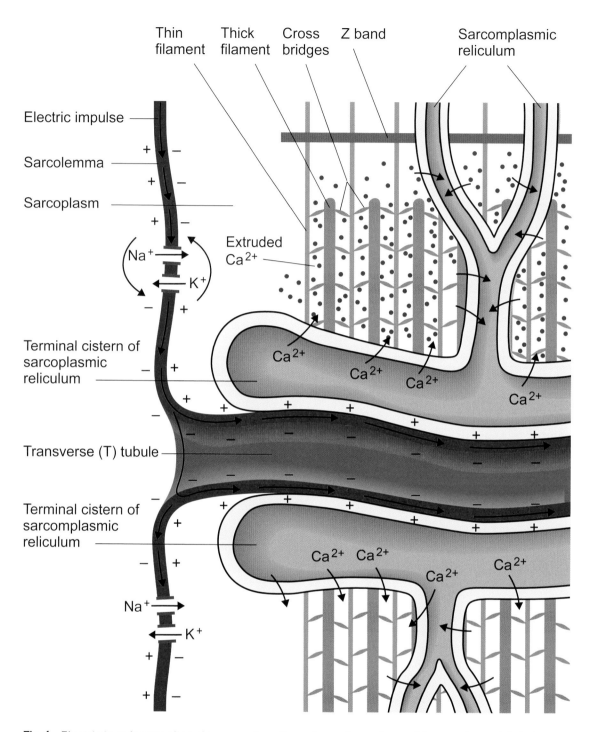

Thin filament   Thick filament   Cross bridges   Z band   Sarcomplasmic reticulum

Electric impulse

Sarcolemma

Sarcoplasm

Na^+

K^+

Extruded Ca^{2+}

Ca^{2+}

Ca^{2+}

Ca^{2+}

Ca^{2+}

Terminal cistern of sarcoplasmic reliculum

Transverse (T) tubule

Terminal cistern of sarcomplasmic reliculum

Ca^{2+} Ca^{2+} Ca^{2+} Ca^{2+}

Na^+

K^+

Fig. 6. Electric impulse traveling along muscle cell membrane (sarcolemma) from motor endplate (neuro-muscular junction) and then along transverse tubules affects sarcoplasmic reticulum causing extrusion of Ca^{2+} to initiate contraction by "rowing" action of cross bridges, sliding filaments past one another.

In more recent studies small intact animal muscles, e. g. soleus (Clausen & Kohn, 1977), extensor digitorum longus (EDL), extensor digitorum brevis, epitrochlearis and sterno-hyoideus are used (Mu et al. 2011). Quite recently, bundles of muscle fibers were isolated from abdominal rectus in patients undergoing abdominal surgery. These preparations weighed around 600 mg and in Krebs-Ringer buffer at 30 °C turned out to maintain force (2 s stimulation at 60 Hz) for at least 12 h in vitro (Skov et al., 2015). The standard conditions for incubation of isolated muscles are in Krebs-Ringer bicarbonate buffer (KR) at room temperature or 30-37 °C during continuous gassing with a mixture of 5 % CO_2 and 95 % of O_2, allowing the maintenance of a constant pH of 7.4. As already mentioned the incubation may also take place at 30 °C in air without buffer allowing the maintenance of force and Na^+- K^+-contents comparable to those in KR buffer (Clausen, 2008a+b). The isotopic tracers ^{22}Na, ^{24}Na, ^{42}K, ^{86}Rb and flame photometry were used to quantify the fluxes of Na^+ and K^+ in µmoles per g wet wt. The intracellular accumulation of the isotopes may be followed over time intervals from seconds to hours if followed by a 4x15 min washout in ice-cold buffer allowing washout of extracellular isotopes with correction for loss of intracellular isotope (Buchanan et al., 2002). ^{42}K and ^{86}Rb are both used as isotopic tracers for K^+, and as shown in Fig. 5, the Na^+, K^+ pump mediated (ouabain-suppressible) uptake of these two tracers in intact muscles are closely similar, both when stimulated by insulin, salbutamol, CGRP and when inhibited by graded concentrations of ouabain (Dørup & Clausen, 1994). Because the half-life of ^{42}K is very short (12.5 hrs), ^{86}Rb is often preferred. However, the efflux (fractional loss) of ^{86}Rb

from rat soleus is 2.3 times smaller than that of ^{42}K. Salbutamol and CGRP increase ^{86}Rb efflux, but inhibit ^{42}K efflux. Finally, the inhibitory effect of bumetanide on ^{86}Rb uptake gives rise to the false impression that skeletal muscle contains a $NaKCl_2$ co-transport system. This implies that for studies of K^+ efflux and bumetanide-sensitive K^+ transport, ^{86}Rb is not even an acceptable tracer for the detection of qualitative changes in skeletal muscle. Control experiments with ^{42}K are essential in any accurate characterization of unknown K^+ transport processes in skeletal muscle.

The transmembrane Na^+, K^+ concentration gradients are generated by the continuous operation of the Na^+, K^+ pumps, moving Na^+ ions actively out of the cell (in Fig. 4 from right to left) and K^+ ions actively into the cell (in Fig. 4 from left to right). By far the major passive influx pathways for Na^+ are the voltage sensitive Na^+ channels. The concentration gradient for Na^+ drives the symport of amino acids into the cell, the anti-port of Ca^{2+} out of the cell (from right to left), the anti-port of H^+ ions out of the cell and the sym-port of NaCl into the cell. Passive K^+ fluxes are mediated by the voltage-sensitive inward rectifier E_M (inw.), the delayed rectifier E_M (del.), the Ca^{2+}-sensitive and the ATP-sensitive K^+ channels.

The abovementioned Na^+ channels may be blocked selectively by saxitoxin or tetrodotoxin at micromolar concentrations. The high affinity of these agents for the Na^+ channels allow them (when labelled with ^3H) to be used for the quantification of the Na^+ channels in nerves or muscles (Harrison et al., 1997). Such measurements are performed by incubating intact skeletal muscles or strips of muscles weighing 20-40 mg at 4 °C for 120 min in buffer containing $2.5x10^{-8}$

M [^{3}H]saxitoxin (0.05µCi/ml) (sufficient to give complete saturation of the specific saxitoxin binding sites), (Harrison et al. 1997; Harrison & Clausen, 1998).

In conclusion, several intact muscle preparations have become available for detailed characterization of Na⁺, K⁺-transport as well as other transport systems in animals and human subjects, both in vitro and in vivo. This has allowed the identification of numerous regulatory anomalies and offered new procedures for compensatory treatment.

►► **Fig. 7.** Diagram describing the sequence of events in excitation-induced redistribution of Na⁺, K⁺ and Cl⁻ in skeletal muscle with values for $[K^+]_o$ and $[Cl^-]_i$ induced by electrical stimulation for 300 s at 10 Hz and published in J. Gen. Physiol. Jan 14, 2013. (a) Na⁺ influx inducing depolarization and increased $[Na^+]_i$. (b) The depolarization activates K⁺ efflux, increasing $[K^+]_o$ from 4 to 42 mM. (c) The increases in $[Na^+]_i$ and $[K^+]_o$ activate the Na⁺, K⁺ pumps. (d) This is a major mechanism for the clearing of extracellular K⁺. (e) The increase in $[K^+]_o$ induces depolarization. (f) This is counterbalanced by the electrogenic action of the Na⁺, K⁺ pumps (g) Elevated $[K^+]_o$ favours vasodilatation, improving the clearance of $[K^+]_o$ via the capillaries (Lo et al., 2004; Armstrong et al., 2007). (h) As the result of depolarization, Cl⁻ influx is promoted, inducing repolarization and increase in $[Cl^-]_i$ from 15 to 38 mM (data from Table 5 in J. Gen. Physiol. Jan. 14, 2013). (i) As the result of this repolarization, K⁺ influx via inward rectifiers is increased. (j) This favours the clearance of $[K^+]_o$.

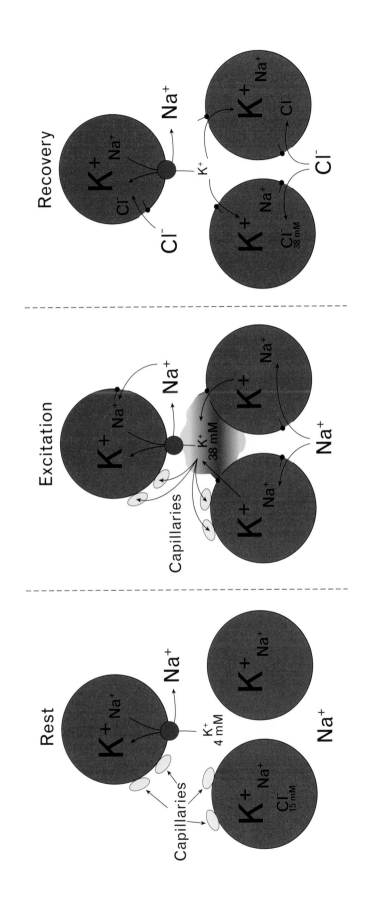

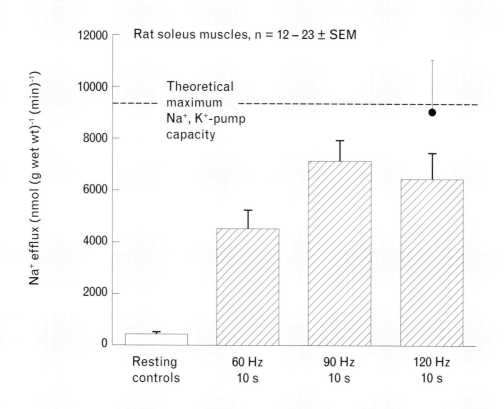

Fig. 8. Na$^+$ efflux in resting (empty column) and electrically stimulated muscles contracting without exerting external force. The Na$^+$ extrusion in stimulated muscles was estimated from the net changes in Na$^+$ content during the first 50 s (cross-hatched columns) or the first 30 s (filled symbol) of recovery from 10 s of stimulation at 60, 90 or 120 Hz (as indicated). Each column or point represents a mean value with bars denoting S.E.M. obtained from 12-23 muscles. (Nielsen and Clausen, J. Physiol.1997). Reproduced with permission from Fig. 5 in Nielsen & Clausen 1997. For details see the same reference.

►► **Fig. 9.** Relationship between the rate of excitation-induced increase in $[K^+]_o$ and the rate of force decline in isolated rat muscles. Soleus and extensor digitorum longus (EDL) muscles were prepared from 4-wk-old rats, mounted for isometric contractions in force transducers and incubated in KR buffer at 30 °C. The muscles were exposed to direct continuous stimulation at frequencies 20-200 Hz, 1 ms pulses at 10 V. The rate of force decline was recorded and expressed as per cent per sec of the maximum force recorded within 0.5 – 1.5 sec after the onset of stimulation. Each point represents the mean ± SEM of observations on 7-16 soleus or 4-21 EDL muscles. (Clausen, Physiol. Reports, 2015).

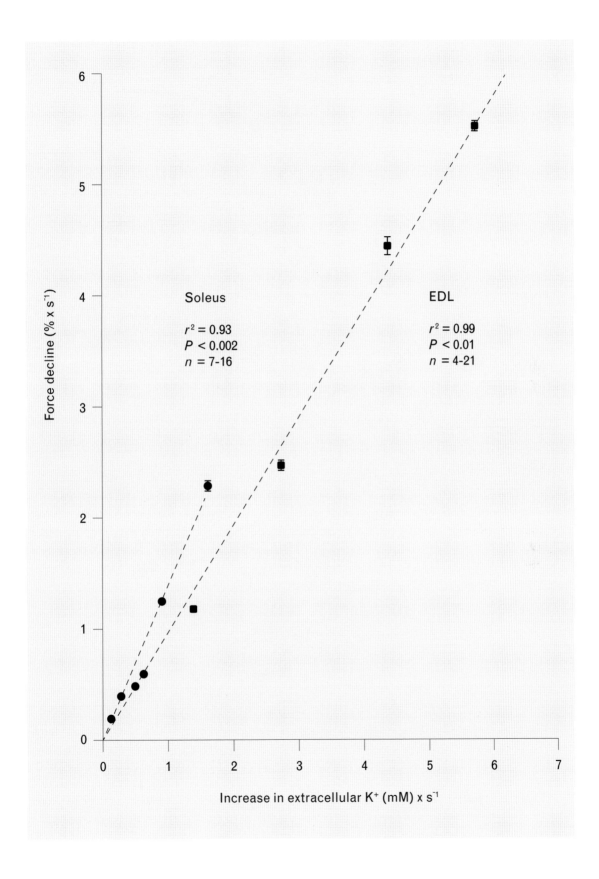

Soleus

$r^2 = 0.93$
$P < 0.002$
$n = 7\text{-}16$

EDL

$r^2 = 0.99$
$P < 0.01$
$n = 4\text{-}21$

Force decline (% x s^{-1})

Increase in extracellular K$^+$ (mM) x s^{-1}

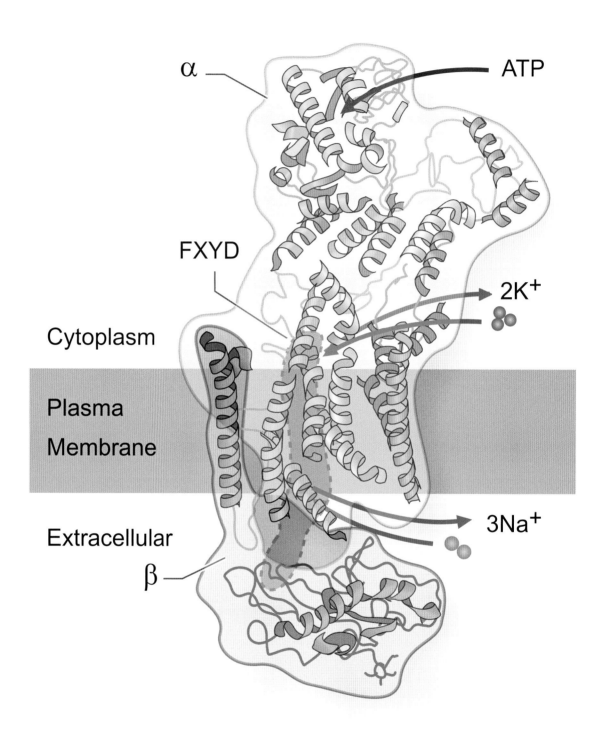

Fig. 10. Model of the Na$^+$, K$^+$ pump by F. Cornelius (copyright) (Shinoda, T., Ogawa, H., Cornelius, F. and Toyoshima, C). If you want to see a simulation of how the Na$^+$, K$^+$-pump molecule works, see the Internet link https://youtu.be/MRKgbwl8vCY.

The definition, structure and function of the Na$^+$, K$^+$-pump

The Na$^+$- K$^+$-pump is a protein molecule with enzyme activity (Na$^+$, K$^+$-ATPase), the property allowing it to be identified and detected in the plasma membrane. The functional definition of the Na$^+$, K$^+$-pump is that it can split the energy-rich molecule ATP of the cells in such a way that its energy is made available to allow active transport of Na$^+$ ions from the cell interior across the cell membrane out into the surroundings of the cell. This outward pumping of Na$^+$ is closely coupled to a simultaneous inward pumping of K$^+$ ions into the cells. Both ions have one positive charge, denoted by a +.

The molecular structure of the Na$^+$- K$^+$-pump is illustrated in Fig. 10 (Shinoda et al., 2009). The entire pump-molecule consists of 3 peptides, the alpha (α) subunit, which performs the actual transport of Na$^+$ and K$^+$, the beta (β) subunit, which moves the entire pump molecule from the ribosomes in the cell, where it is synthesized and transferred into the cell membrane to be situated in such a way that it can perform the pumping of Na$^+$ and K$^+$. This starts by 3 Na$^+$ ions from the cell interior getting access to and occluded into a small cavity inside the pump molecule from where the Na$^+$ ions can be pumped out through the pump molecule in such a way that they can leave the cell. Now the 3 Na$^+$ ions are rapidly replaced by 2 K$^+$ ions from outside the cell entering the same small cavity, where the 3 Na$^+$ ions had been occluded (Glynn, 2002). This process requires energy made available by the splitting of ATP. This very energy-rich molecule becomes bound to the pump at a catalytic site which promotes the splitting of ATP (see α in Fig. 10). Thus, the Na$^+$, K$^+$ pump is defined by its ability to split ATP and utilize the energy thus made available for the active transport of Na$^+$ and K$^+$. The third peptide in the Na$^+$, K$^+$-ATPase, the gamma (γ) subunit, which also has another name (FXYD1), mediates the signals regulating the activity of the Na$^+$, K$^+$ pump (Shattock, 2009). At the body temperature of 37 °C, each Na$^+$- K$^+$-pump molecule splits 8,000 ATP molecules every minute, moves 3x8,000 Na$^+$ ions out of the cell against 2x8,000 K$^+$ ions into the cell. This leads to a net export from the cell of 8,000 positive charges every min per Na^+, K^+-pump. Due to this so-called "electrogenic action", the Na^+,K^+ pumps generate or maintain an electric gradient across the cell membrane, where the cell interior is negatively charged in relation to the cell surroundings. Thus, the Na^+,K^+ pumps aid to the resting membrane potential of the cell by a few mV (2-15 mV negative inside). Indeed, the resting membrane potential itself also relays indirectly on the Na^+,K^+ pumps, which establish the Na^+- and K^+-gradients that, due to the much larger permeability for K^+ than for Na^+, set up a diffusional potential, the resting membrane potential.

In conclusion, the major functions of the Na^+, K^+ pumps are to maintain not only the uneven distribution of Na^+ and K^+ concentrations across the plasma membrane, but also this electric gradient, the membrane potential, which is essential for the excitability and contractility of the muscle cells. A Youtube simulation of Na^+, K^+-pump sequence according to Garlandscience will be shown in the electronic version of this book, see the still illustration in Fig. 10a and the Internet link https://youtu.be/MRKgbwl8vCY.

The synthesis and localization of the Na$^+$, K$^+$ pumps in skeletal muscle

The Na$^+$, K$^+$ pump molecules are synthesized in the ribosomes and then translocated into the plasma membrane (sarcolemma) around the cell as well as in the T-tubular walls. This process is mediated by the β-subunit of the Na$^+$- K$^+$-pump (see Fig. 10). The α-subunit is located in the plasma membrane in such a way, that its binding site for cardiac glycosides is accessible from the extracellular space and t-tubular lumen, available for reversible binding. As shown in Fig. 1 in (Clausen & Hansen, 1974), this binding is saturable, allowing accurate quantification of the content of Na$^+$, K$^+$-ATPase molecules in skeletal muscle. The Na$^+$, K$^+$-ATPase α$_2$ subunit isoform is expressed during development when the t-tubules form (Cougnon et al., 2002). Histochemical analysis show that the

α$_2$ subunit is the sole Na$^+$,K$^+$-ATPase subunit localized to the t-tubular membranes, whereas the α$_1$ subnunit is localized to the surface sarcolemma (Williams et al., 2001). The α$_2$ subunit comprise up to 90 % of total α subunit content and has a lower affinity for K$^+$ than α$_1$ (Orlowski & Lingrel, 1988). This lower affinity for K$^+$ extends the regulatory range for stimulation of α$_2$ activity by K$^+$ site occupancy, allowing it to respond to T-tubular lumen (extracellular) K$^+$ concentrations up to 40 mM. This was recently proposed to provide a rapidly recruitable reserve mechanism for increasing α$_2$ activity in working muscles (DiFranco et al., 2015).

In conclusion, in skeletal muscle, the distribution of Na$^+$ and K$^+$ ions as well as their transport rates depend on the content, subtypes and localization of Na$^+$, K$^+$-pumps.

06

Measurements of the content of [³H]ouabain binding sites indicate that in skeletal muscle, the major part of these binding sites of the Na⁺, K⁺ pumps are localized on the outer surface of sarcolemma and in the t-tubules

The first measurements of [³H]ouabain binding to intact mammalian muscles showed that virtually all these high-affinity binding sites are situated on the outer surface of the muscle cells, where binding reaches equilibrium within 120 min when a concentration of ouabain in the incubation medium of 10^{-6} M is used (Clausen & Hansen, 1974). The binding of [³H]ouabain is slowly reversible, in particular at 0 °C, allowing the "trapping" of a substantial fraction of the [³H]ouabain bound to the intact muscles within 120 min of incubation at 30 °C, followed by 4x30 min of washout at 0 °C. The binding of [³H]ouabain may also take place in vivo, following an intraperitoneal injection of saturating doses of [³H]ouabain. Fig. 2 in Clausen et al., 1982 shows the time-course of the binding of [³H]ouabain

to rat soleus and EDL muscles following an intraperitoneal injection of increasing doses of [³H]ouabain (0.75 – 20 nmoles/g body wt.). The binding values of soleus and EDL muscles (pmoles/g wet wt.) are the same in vitro and in vivo, provided sufficient concentration of [³H]ouabain and duration of binding are used (Table 3 in Clausen et al., 1982). In isolated soleus muscle which had reached almost maximum binding of ³H ouabain at 30 °C, the addition of unlabeled ouabain caused a prompt and four- to fivefold rate of release of ³H-ouabain from the muscle, most likely to reflect displacement of ³H ouabain from binding sites at the outer surface of the cells (Clausen et al., 1982). In contrast, making the muscle cells leaky by transverse cuts caused no loss of ³H-activity, showing that there was no intracellular accumulation of free diffusible [³H]ouabain (Clausen & Hansen, 1974).

A striking feature of skeletal muscle is the intracellular network of tubules (the tranverse or t-tubules) (Fig. 11) which connects the surface sarcolemma with the cell interior (Peachey & Eisenberg, 1978). This highly branched network facilitates the access of extracellular molecules to the cell interior by allowing diffusion of substances from the surroundings of the cell and into the tubular lumen. Moreover, the network augments the surface area of the plasma membrane separating the extracellular space from the intracellular space. Thus, Na^+, K^+, Cl^- and Ca^{2+} ions may reach the cytoplasm not only via the surface plasma membrane, but also via the much larger area of the t-tubule membranes (in frog sartorius muscle 7-fold larger than that of the outer sarcolemma which has been shown to contain the necessary specific ion channels (Peachey & Eisenberg, 1978)). 70-90 % of the inward rectifier K^+ channels reside in the t-tubular membranes. In the skeletal muscles of adult rats, the ratio beween the area of t-tubular membranes and that of the sarcolemma varies from 3 to 5 (Appelt et al., 1989). As mentioned above, the t-tubular membranes also contain Na^+, K^+-pumps, predominantly the α_2 subunit. This implies that during excitation, the ionic composition of the tiny volume of the t-tubular lumen is likely to be dominated by the fluxes via the many Na^+ and K^+ channels present in the tubular walls. It has been calculated that for each action potential in frog skeletal muscle, the K^+ concentration in the t-tubular lumen increases by 0.37 mM

(Kirsch et al., 1977). Thus, during stimulation at 40 Hz, the K^+ concentration in the tubular lumen may increase by around 15 mM in 1 s (40x0.37 mM), sufficient to block further impulse propagation. Others have reported similar estimates (Almers, 1980). The pathways of the K^+ causing repolarization of the t-tubules are the K^+ channels in the extensive t-tubular walls, and unless this K^+ is returned to the cytoplasm via the Na^+, K^+-pumps in the t-tubules, it has to be cleared into the surroundings of the cell. Because this has to take place via the narrow lumen of the relatively long and narrow t tubules, however, the clearance of K^+ from the t-tubular lumen is likely to be considerably delayed (Kirsch et al., 1977) and to a large extent will depend on reaccumulation of K^+ via the Na^+, K^+ pumps into the cytoplasm.

In conclusion, due to the small volume of t-tubular lumen and the vast number of surrounding transporters, the Na^+, K^+ turnover in this narrow space must be exceedingly fast during excitation. Therefore, the maintenance of stable concentrations of Na^+ and K^+ and the excitability of the t-tubular membranes depend on a very close match between the rates of Na^+, K^+ transport via the Na^+ channels, the K^+ channels and the Na^+, K^+ pumps. The importance of K^+ concentration changes for the action potentials and membrane currents has been analyzed by computer simulation (Wallinga et al., 1999). However, the analysis might well have underestimated the role of the Na^+, K^+ pump capacity per unit area in the t-tubules, which was assumed to be one-tenth of that of the sarcolemma.

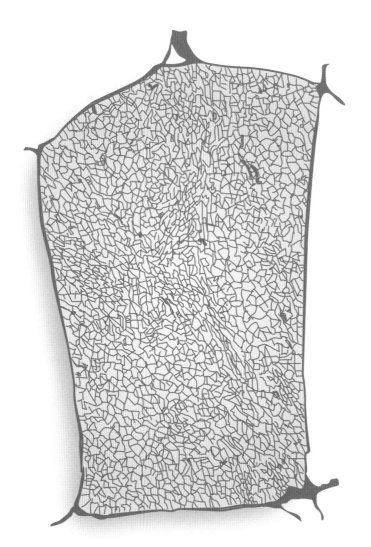

◄ **Fig. 11.** Reconstruction of T system of frog twitch muscle fiber made by tracing T tubules stained with peroxidase method in series of high-voltage electron micrographs of serial transverse sections 0.7μm thick. Tubules projected into transverse plane of fiber. 1,000 kV; x1,800. (Peachey and Eisenberg, 1978, by copyright permission of the Biophysical Society).

►►▲ **Fig. 12.** Diagrammatic representation of a muscle biopsy during incubation with [3H]-ouabain and vanadate. The open ends of the muscle cells allow vanadate (VO_4^{---}) to gain access to the phosphorylation site of the Na$^+$, K$^+$ pump on the inner surface of the plasma membrane. The high-affinity binding of vanadate maintains the Na$^+$, K$^+$ pump in a configuration capable of binding [3H]-ouabain to the outer surface of the Na$^+$, K$^+$-pump molecule. (Clausen & Nielsen, CNS, 1999).

►►▼ **Fig. 13.** Relative changes in the content of Na$^+$, K$^+$ pumps in human vastus lateralis muscle associated with some physiological and pathophysiological conditions. All values, except one, are based on determination of the total content of [3H]ouabain binding sites in Bergström needle biopies. The value given for muscular dystrophy was measured in muscle cells cultured from patients. Each condition and the reference are given on the left side of the diagram and the percentage increase or decrease is expressed in relation to the values obtained in parallel measurements on control subjects. (Clausen, 1998, copyright the Biochemical Society and the Medical Research Society)

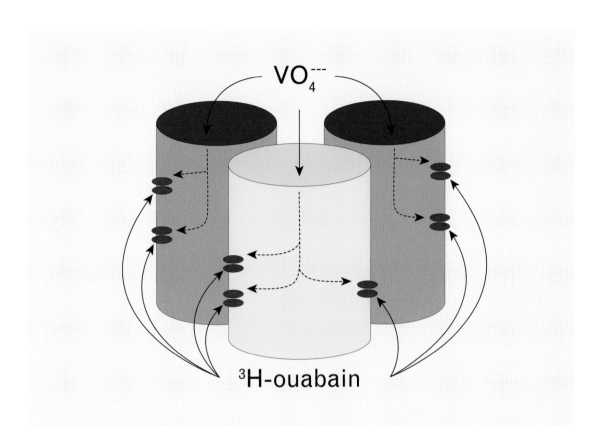

VO_4^{---}

^3H-ouabain

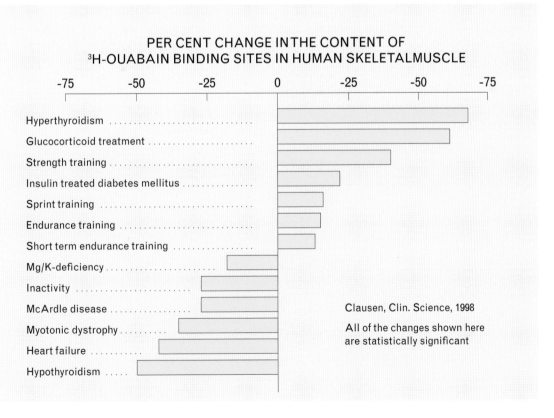

PER CENT CHANGE IN THE CONTENT OF
^3H-OUABAIN BINDING SITES IN HUMAN SKELETALMUSCLE

Clausen, Clin. Science, 1998

All of the changes shown here
are statistically significant

The content of Na$^+$, K$^+$ pumps in skeletal muscles and how it can be quantified

Even though we often speak about the Na$^+$, K$^+$ pump in singular, we must not forget that most tissues contain a very large number of Na$^+$, K$^+$ pump molecules. Thus, measurements on many samples of human skeletal muscles show that each gram contains around 300 pmoles Na$^+$, K$^+$ pumps (300x10^{-12} moles). This content in molar units can be converted to number of Na$^+$, K$^+$ pump molecules by multiplying with the number of molecules per mole (6×10^{23}): 300x10$^{-12} \times 10^{23} \times 6 = 1800 \times 10^{11} = 180 \times 10^{12}$. This corresponds to about 180 billions Na$^+$, K$^+$ pump molecules. (one billion corresponds to $10^6 \times 10^6 = 10^{12}$). This astronomical number sounds excessive, but it should be recalled that most of these Na$^+$, K$^+$ pumps are localized in the very large surface area of the t-tubules (see section 6, fig 6 and 11). Each Na$^+$, K$^+$ pump molecule comprises an alpha (α), a beta (β) and a gamma (γ) subunit with a total molecular weight of 147,000. This means that each gram wet wt. of muscle contains 44 µg Na$^+$, K$^+$ pumps.

Several plants produce poisonous compounds (digitalis glycosides) that specifically bind to and inhibit the Na$^+$,K$^+$-ATPase. The binding site is similar for these compounds and evolutionarily conserved (Lingrel, 2010). The molecular structure of the binding site of ouabain, a digitalis glycoside, was published in 2009 (Ogawa et al., 2009), demonstrating that ouabain binds deeply into the transmembrane domain of the Na$^+$, K$^+$ATPase α-subunit. These digitalis glycosides can be labelled with radioactive tritium (^{3}H) allowing the measurement of the labelled molecules in a scintillation counter. Thus the quantification of Na$^+$, K$^+$ pumps in a muscle biopsy weighing 5 mg can be done by incubating groups of 3-4 biopsies for 120 min at 37 °C in a few milli-litres of a tris-sucrose buffer containing 10^{-6} M of the [^{3}H] labelled compound (most often the water soluble digitalis glycoside ouabain but also the less soluble cardiac glycoside digoxin). Human muscle biopsies bind the same amount of ^{3}H-ouabain per g muscle wet wt. as ^{3}H-digoxin (Nørgaard et al., 1984). The [^{3}H]labelled ouabain molecules will then bind firmly and selectively to the Na$^+$ K$^+$ pump molecules, one molecule to each Na^+, K^+ pump molecule – a so-called stoichiometric binding. This requires that the buffer contains the phosphate analog vanadate (VO_4), which via the cut ends of the muscle biopsies enters the cell interior (see the diagram of cut muscle cells in Fig. 12) and gains access to the inner surface of the Na^+, K^+ pumps and as first demonstrated by Otto Hansen (Hansen et al., 1979) facilitates the binding of [³H]ouabain to the outer surface of the sarcolemma (muscle cell membrane).

After incubation, the 5 mg muscle biopsies are removed from the incubation tubes and washed in ice-cold buffer to remove the [³H]ouabain not bound to the surface membranes of the muscle cells. The biopsies are then blotted on dry filter paper, weighed and transferred into counting vials containing a liquid scintillation mixture. The [³H]ouabain bound to the biopsy is measured by counting the [³H] activity in a scintillation counter and comparing the value to the [³H]ouabain concentration in the incubation buffer. On the basis of the specific activity of the [³H]ouabain in the buffer it is possible to express the [³H]ouabain bound to each muscle biopsy in pico-moles (10^{-12} moles) per g wet wt.

Such measurements can be performed on 50-100 muscle biopsies in one thermostated incubation bath in a few hours. This simple and economic method has for 30 years been used by many scientists and technicians around the world, and as shown in Table 1 there is good agreement between the results obtained in 25 studies on biopsies of human vastus lateralis muscle performed on control subjects in several different laboratories (Nørgaard et al. 1983; Clausen, 2003). The average value is 289 pmol/g wet wt. with modest scatter in each group of data.

References	Content, pmol/g wet wt
Nørgaard et al., 1984	278 ± 15
Dørup et al., 1988	258 ± 16
Kjeldsen & Grøn, 1989	268 ± 17
Klitgaard & Clausen, 1989	276 ± 19
McKenna et al., 1993	333 ± 19
Green et al., 1993	339 ± 16
Schmidt et al., 1993	257 ± 28
Madsen et al., 1994	307 ± 43
Leivseth & Reikeras, 1994	306 ± 27
Schmidt et al., 1994	223 ± 13
Gullestad et al.,1995	258 ± 13
Evertsen et al., 1997	343 ± 11 (men)
Evertsen et al., 1997	281 ± 14 (women)
Haller et al., 1998	317 ± 37
Green et al., 1999	289 ± 22
Green et al., 2000a	348 ± 12
Green et al., 2001	268 ± 19 (men)
Green et al., 2001	243 ± 13 (women)
Aughey et al. 2005	307 ± 41
Nordsborg et al., 2005	326 ± 30
Green et al., 2009	259 ± 15
Petersen et al., 2011	280 ± 50
McKenna et al., 2012	353 ± 76
Boon et al., 2012	240 ± 10
Goodman et al., 2014	250 ± 60

It is important that the method can be used for frozen samples of skeletal muscle, allowing long-distance transport with dry ice by flight without any loss of the ability to bind ^{3}Houabain during incubation during the standard incubation at 37°. Moreover, long-term storage of the frozen samples has been shown to be possible for at least 4 years (Clausen, 2003). The VO_4-facilitated [^{3}H]ouabain binding assay has also been adapted for measurements on specimens from the human heart (Nørgaard et al., 1988; Nørgaard et al., 1990), porcine and canine heart (Schmidt et al., 1990) and uterine smooth muscle (Everts et al., 1990). A comparison of the binding kinetics of [^{3}H]ouabain in a variety of human tissues showed that the affinity of α_1, α_2 and α_3 subunit isoforms for ouabain is similar (Müller-Ehmsen et al. 2001; Wang et al. 2001). This confirms repeated observations that in biopsies of human skeletal muscle and myocardium, only a single population of high-affinity sites can be detected (Kjeldsen, 1986a+b; Nørgaard, Kjeldsen & Clausen, 1984; Nørgaard, Kjeldsen & Hansen, 1984). This implies that in human skeletal muscle and myocardium, Na$^+$, K$^+$-pumps can be quantified by measuring the total content of [^{3}H]ouabain binding sites.

◄◄ **Table 1:** The contents of ^{3}H-ouabain binding sites in human skeletal muscle biopsies as determined in 25 different studies in the 30-year time-interval 1984-2014. References, 1st author and year of publication, also given in the total list of references at the end of the general text. Contents ± SEM (pmoles/g wet wt.).

A review of measurements of the contents of ^{3}H-ouabain binding sites in picomoles per g wet wt (Clausen, 1998) showed that the contents of Na$^+$, K$^+$ pumps in human skeletal muscle associated with physiological and pathophysiological conditions undergo increases from 70 % to decreases around 50 % (Fig. 13). All of the changes shown in that figure are statistically significant.

The measurement of [^{3}H]ouabain binding raises the question whether this quantifies all the functional Na$^+$, K$^+$ pumps present in the intact muscle. Is it possible to induce a rapid activation of all the Na$^+$, K$^+$ pumps in the muscle? In rat soleus, electric stimulation at 120 Hz for 10 s increases intracellular Na$^+$ to around 50 mM (Nielsen & Clausen, 1997). When these muscles were subsequently allowed to rest in standard KR at 30 °C, the net efflux of Na$^+$ measured over the first 30 s was shown to reach 9,000 nmol/g wet wt./min, 97 % of the theoretical maximum Na$^+$ efflux rate of 9,300 nmol/g wet wt/min (Fig. 5 in Nielsen & Clausen, 1997). Measurement of maximum binding capacity

for [^{3}H]ouabain in isolated intact rat soleus muscle (0.72 nmol/g wet wt.) would predict that if all Na$^+$, K$^+$ pumps were operating at full speed, the theoretical maximum K$^+$ uptake should reach 0.72 nmol x 8,900 = 6,408 nmol/g wet wt./min at 30 °C (corrected for temperature using the observed Q_{10} of 2.3 for the rate of active Na$^+$, K$^+$-transport in KR buffer (Clausen & Kohn, 1977). More recent measurements on intact mouse flexor digitorum brevis muscle fibres showed that the α_2 subunit of Na$^+$, K$^+$ATPase has a Q_{10} of 2.1, which is in satisfactory agreement with the earlier value of 2.3 (DiFranco et al., 2015). When rat soleus muscles are loaded with Na$^+$ by preincubation in K$^+$-free KR without Ca^{++} and Mg^{++}, [Na$^+$]$_i$ reaches 126 mM (Clausen et al., 1987). When these muscles are subsequently incubated for 3 min in KR buffer containing ^{42}K or ^{86}Rb and between 5 and 139 mM K$^+$, the ouabain-suppressible rates of ^{42}K or ^{86}Rb uptake can be measured and the results shown in an Eadie-Hofstee plot (see Fig. 4 in Clausen et al., J. Physiol. 1987). The maximum rates of ^{42}K and ^{86}Rb uptake determined from this plot reach 6,150 nmol/g wet wt./min, corresponding to 96 % of the abovementioned theoretical maximum K^+ uptake at 30 °C. Using the same Na^+ loading procedure, the maximum rate of ouabain-suppressible ^{86}Rb uptake was closely correlated (r = 0.95; P<0.001) to the content of [^3H]ouabain binding sites over a wide range of values obtained by varying thyroid status, age, or K^+ depletion (Fig. 5 in Clausen, 2013).

The VO_4-faciliteted ^3H-ouabain binding assay also allows measurements to be performed on muscle samples taken post mortem (Nørgaard et al., 1985). A comparison of the values obtained in fresh rat skeletal muscle with those determined after storage at 20 °C for 12 h showed that the loss of the ability to bind [^3H]ouabain was surprisingly slow, around 1 %/h.

Studies on vastus lateralis muscle samples obtained from 10 human subjects 0.5 – 6 h after death showed that the content of [^3H]ouabain binding sites declined by only 8 % in 6 h (Nørgaard et al., 1985).

In conclusion, [^3H]ouabain binds stoichiometrically to a specific receptor on the α-sub-unit of the $Na^{+,} K^+$-ATPase, allowing accurate quantification of the total content of Na^+, K^+ pumps in intact skeletal muscle, as well as in muscle biopsies weighing around 5 mg. The method can be applied to frozen or post mortem tissue samples with closely similar results to those obtained wit fresh biopsies.

The ^3H-ouabain binding method has allowed the detection of numerous regulatory and pathophysiological changes in human skeletal muscle (Clausen, 1998; Clausen, 2010; Clausen, 2013a). The most important advantage of the method is that the content of Na^+, K^+ pumps it measures represents Na^+, K^+ pumps, which may all very rapidly become operational. This implies that at variance with most other methods, the $^3[H]$ouabain binding assay allows the quantification of the total content of functional Na^+, K^+ pumps in muscle tissue. When activated by intracellular Na^+ loading and increased extracellular K^+, the values quantified are in close agreement with the maximum values for ouabain-suppressible uptake of ^{42}K or ^{86}Rb (for details, see Clausen et al., J. Physiol., 1987; Clausen, J. Gen. Physiol. 2013b).

►► **Fig. 14.** Section 8/page 1 (edl in vitro in vivo, USB 2008) Stimulation-induced expected rise in [K^+]$_o$ in EDL muscles in six different experiments reported by Clausen, Fig. 2 in J. Gen. Physiol., vol. 141, pp. 179-192, 2013. Values are given as columns with bars denoting **SEM** and the number of muscles inside the columns. The frequency and duration of stimulation in vitro and in vivo are given below each column together with the P-values for the statistical significance of the rise in extracellular K^+ obtained in each group of experiments.

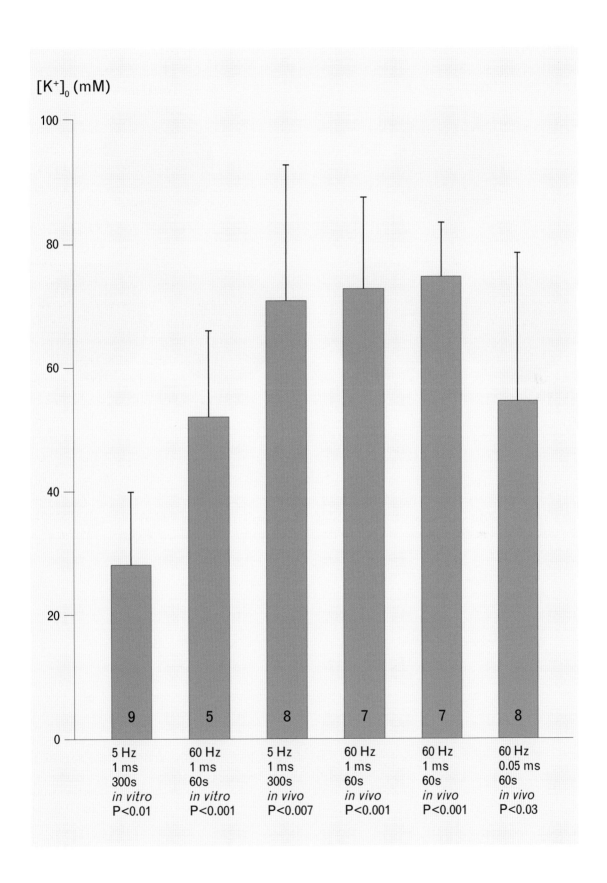

[K+]0 (mM)

5 Hz	60 Hz	5 Hz	60 Hz	60 Hz	60 Hz
1 ms	1 ms	1 ms	1 ms	1 ms	0.05 ms
300s	60s	300s	60s	60s	60s
in vitro	*in vitro*	*in vivo*	*in vivo*	*in vivo*	*in vivo*
P<0.01	P<0.001	P<0.007	P<0.001	P<0.001	P<0.03

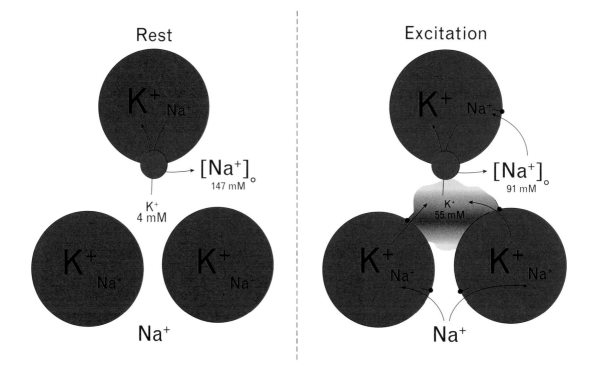

Fig. 15. Diagram of Na^+, K^+ in rat EDL muscle at rest and during electrical stimulation at 60 Hz for 60 sec. The stimulation leads to opening of the Na^+ channels allowing rapid entry of Na^+ ions into the muscle cells, causing depolarization, and rapid loss of K^+ ions via voltage-sensitive K^+ channels. The implication of this redistribution is that in the interstitial water space $[Na^+]_o$ decreases from 147 to 91 mM and $[K^+]_o$ increases from 4 to 55 mM, a combination sufficient to cause rapid decrease in excitability and muscle fatigue. This is restored primarily by acceleration of active electrogenic Na^+, K^+ transport via the Na^+, K^+-pumps, indicated in blue. (Clausen, *Physiological Reports,* 2015).

Muscle contractions, Na^+, K^+ transport and sodium potassium fatigue

The many Na^+-K^+ pumps situated in the cell membranes maintain inside the muscle cells a low concentration of Na^+ and a high concentration of K^+. In rat resting soleus muscles, the passive influx of Na^+ is low (0.85 ± 0.03μmol/g wet wt/min (N = 8) (Clausen & Kohn, 1977), and the energy turnover related to the active Na^+K^+-fluxes in resting muscles is only around 5-10 % of total energy turnover (Clausen et al., 1991). During action potentials triggered by electric pulses at 30 °C, the influx of Na^+ in isolated rat EDL muscle amounts to 12 nmol/action potential/g wet wt. and in isolated rat soleus to 1.9 nmol/action potential/g wet wt. (Clausen et al., 2004). This excitation-induced influx takes place via the Na^+ channels and is immediately followed by an almost equimolar efflux of K^+ (10 nmol/g wet wt. in EDL and 1.6 nmol/g wet wt. in soleus). This rapid efflux of K^+ causes a rise in extracellular K^+, which has repeatedly been proposed as a likely cause of fatigue (Fenn, 1940; Bigland-Ritchie et al., 1979; Jones, 1981; Sjøgaard, 1990; Juel, 1988; Clausen & Everts, 1991; Clausen &

Nielsen, 1994; Balog & Fitts, 1996; Sejersted & Sjøgaard, 2000; Green et al., 2000b; Clausen, 2003a; Clausen, 2003b; Nordsborg et al., 2003; Clausen, 2013b; Clausen, 2015). As shown in Fig. 14, several experiments in vitro and in vivo showed that excitation (5-60 Hz) of rat EDL muscles causes $[K^+]_o$ to reach values between 25 and 75 mM. All of these increases are statistically significant (from P <0.001 to P = 0.03 (Clausen, 2013a)).

More recently, it was observed that during excitation, the almost concomitant decrease in $[Na^+]_o$ in a synergistic way augments the inhibitory action of $[K^+]_o$ (Bouclin et al., 1995; Overgaard et al., 1997; Cairns et al., 2003). Recent studies of ^{22}Na uptake in isolated rat EDL muscles showed that 60 s of 60 Hz stimulation increased the intracellular uptake of ^{22}Na by 12 μmoles/g wet wt. (Clausen, 2015). This amount of ^{22}Na was lost from an extracellular water space of 0.214 ml, (12 μmoles (g wet wt)$^{-1}$/0.214 ml) corresponding to a drop of 56 μmoles(ml)$^{-1}$ or 56 mM Na^+, reaching an average Na^+ concentration of 91 mM (147-56 mM) in the interstitial water space between the muscle cells. This drop in $[Na^+]_o$ is replaced by an almost equimolar amount of K^+ (11 µmoles (g muscle wet wt)$^{-1}$/0.214 ml, increasing $[K^+]_o$ by 51 mM (11 µmol(g wet wt)$^{-1}$/0.214 ml, reaching 4 + 51 mM = 55 mM, similar to the drop in $[Na^+]_o$ of 56 mM. These changes are illustrated in the diagram of muscle cells shown in Fig. 15.

The mechanisms of sodium-potassium fatigue have been explored in detailed studies of the effects of reduced Na^+/K^+ gradients and Na^+, K^+ pump stimulation on compound action potentials (M-waves) in isolated rat soleus muscle stimulated via the nerve (Overgaard et al., 1999). Tetanic force was modulated by continuous nerve stimulation without and with tubocurarine (blocks the acetyl choline activation at the motor end-plate) (or tetrodotoxin, a specific and potent inhibitor of the Na^+ channels). As shown in Fig. 5 in Overgaard et al., 1999, there is a close correlation between excitability evaluated by the M wave area and tetanic force.

In conclusion, these observations indicate that loss of excitability is an important factor in fatigue induced by for example 30 or 60 Hz stimulation. Excitation seems to be a self-limiting process due to the run-down of Na^+, K^+ gradients and depolarization, recoverable by stimulation of the Na^+, K^+ pumps (Nielsen & Clausen, 2000; Clausen, 2003; Clausen, 2015).

The rate of [³H]ouabain binding to sarcolemma can be quantified and what information may this provide?

The rate of [³H]ouabain binding to isolated skeletal muscle can be measured by short-lasting incubation (1-15 min) at 30 °C in KR buffer containing [³H]ouabain (2 µCi/ml with a subsaturating concentration of ouabain (10^{-7} M)) followed by immediate wash for 4x30 min in ice-cold KR buffer to remove the [³H]ouabain remaining in the extracellular space, allowing the [³H]ouabain already bound to its receptors to remain bound (Everts & Clausen, 1994). As shown in Fig. 4 in Clausen & Hansen, 1977, in isolated rat soleus muscle, the rate of [³H]ouabain binding is closely correlated to the ouabain-suppressible ^{42}K influx. It turns out that in skeletal muscle preparations, the rate of [³H]ouabain binding increases linearly with the content of [³H]ouabain binding sites per g wet wt. as well as with the rate of active Na^+, K^+-transport. E.g., the stimulating effect of insulin-like growth factor (IGF-I) is associated with an increased rate of [³H]ouabain binding (Dørup & Clausen, 1995). Like insulin, CGRP, and epinephrine, amylin also increases the rate of [³H]ouabain binding without causing any change in the total content of [³H]ouabain binding sites in rat soleus and EDL (Clausen, 2000). The effects of insulin and IGF-I on ^{86}Rb uptake and $[Na^+]_i$ are not additive, indicating that these two peptides act via the same mechanism, presumably by increasing the frequency by which the Na^+-K^+-pump molecules pass through the configuration optimal for binding the ^3Houabain molecules.

In conclusion, the stimulating action of IGF-I on the rate of [³H]ouabain binding is a specific indication that this hormone acts on the rate of Na^+, K^+ pumping. Thus, measurements of the rate of [³H]ouabain binding seems to provide a rather selective demonstration of the participation of the Na^+, K^+ pumps in Na^+, K^+ transport of isolated muscle tissue (Clausen & Hansen, 1977).

Measurements of [^{3}H]ouabain binding to other cell types and preparations

For sake of comparison with the binding to isolated muscles, the binding of [^{3}H]ouabain to a number of other tissues and cells were measured. In the myocardium of rats, the affinity of the [^{3}H]ouabain binding sites was clearly lower than in skeletal muscle, preventing accurate quantification of the content of [^{3}H]ouabain binding sites. In contrast, biopsies from the myocardium of guinea pigs and human subjects show a high affinity for [^{3}H]ouabain binding and has allowed quantification of [^{3}H]ouabain binding sites in several studies (Buur et al., 1982; Clausen 1998).

As shown in Table 2, measurements of [^{3}H]ouabain binding sites in several human tissues show an extensive hierarchy of different levels, ranging from 0.071 ± 0.0014 pmoles/ml in the erythrocytes to 11,351 ± 177 pmoles/g wet wt. in the brain cortex (Schmidt et al., 1992). This almost 160,000-fold range of capacity for active Na$^+$K$^+$ transport fits with the marked differences in the need for keeping pace with the passive leaks for Na$^+$ and K$^+$. Obviously these fluxes are much larger in tissues like the brain cortex generating many more action potentials. Another study showed that in patients with dementia, the content of ^{3}H-ouabain binding sites in frontal cortical cortex was 4274 ± 1020 whereas in controls it was 11 397 ± 976 pmol/g wet wt. (Schmidt et al., 1996). Conversely, in smooth muscle preparations obtained from biopsies of human uterus, the content of of [^{3}H]ouabain binding sites was only 83 pmol/g wet wt, which is 137-fold smaller than in normal human brain cortex (Everts, Skajaa & Hansen, 1990). In intact rat sciatic nerves, [^{3}H]ouabain is bound with high affinity allowing quantification in the range 145 – 170 pmol/g wet wt. (Kjeldsen & Nørgaard, 1987). The human myocardium contains about twice the concentration of ^{3}H-ouabain binding sites measured in the biopsies from vastus lateralis muscle (table 2). The content of ^{3}H-ouabain binding sites in lymphocytes, monocytes and leucocytes is clearly higher than that of the erythrocytes, reflecting their much larger Na$^+$, K$^+$-pump activity. At variance with the data in Table

1, the pronounced differences between those in table 2 demonstrates that the vanadate-facilitated ^{3}H-ouabain binding site method allows the detection of large functional variations in Na$^+$,K$^+$-pump capacity between tissues and cell types.

In conclusion, measurement of ^{3}H-ouabain binding capacity allows the quantification of the contents of Na$^+$, K$^+$-pumps in a wide variety of tissues and free cells and to relate them to species and Na$^+$, K$^+$-pump activity.

Tissue or cell type	Concentration of Na$^+$, K$^+$-pumps
Erythrocytes	0.071±0.0014 (pmol/ml)
Lymphocytes	3.5±0.8 (pmol/ml)
Monocytes	4.1 (pmol/ml)
Leucocytes	1.9±0.5 (pmol/ml)
Uterine smooth muscle	83±9 (pmol/g)
Heart, endomyocardium	559±62 (pmol/g)
Heart, endomyocardium	505±41 (pmol/g)
Myocardium, left ventricle	760±58 (pmol/g)
Myocardium, left ventricle	728±58 (pmol/g)
Myocardium homogenate	507±21 (pmol/g)
Brain cortex	11351±177 (pmol/g)

Table 2 Concentration of Na$^+$,K$^+$-pumps in human cells and tissues determined from the ^{3}H-ouabain binding capacity (T. Clausen, Clinical Science, 1998, 95, 3-17). Values are expressed as pmol per g wet wt. or per ml cells, and are means ± S.E.M. As the wet weight of 1 ml of cells is not much above 1 g, all values given are readily comparable.

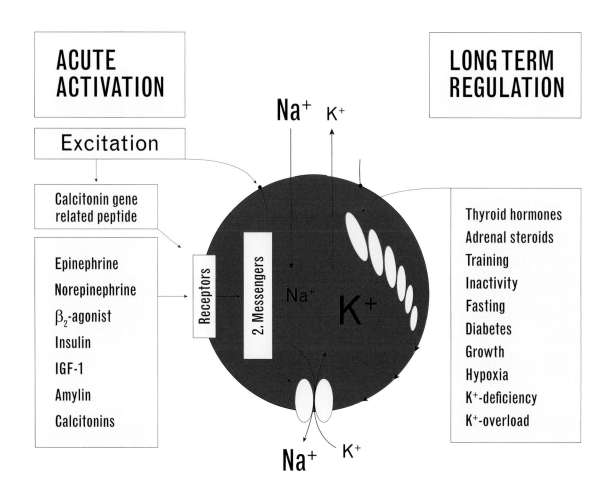

Fig. 16. Diagram of regulatory factors controlling the activity and contents of Na$^+$, K$^+$pumps in skeletal muscle. *Left:* factors eliciting acute stimulation of the Na$^+$, K$^+$ pump activity via receptors and second messengers. *Right:* factors influencing the content of Na$^+$, K$^+$ pumps by modifying their synthesis or degradation. IGF-I, Insulin-like growth factor I. The Na$^+$, K$^+$-pump molecule is shown in the bottom of the muscle cell (Modified from Clausen, *Physiol Reviews*, 2003).

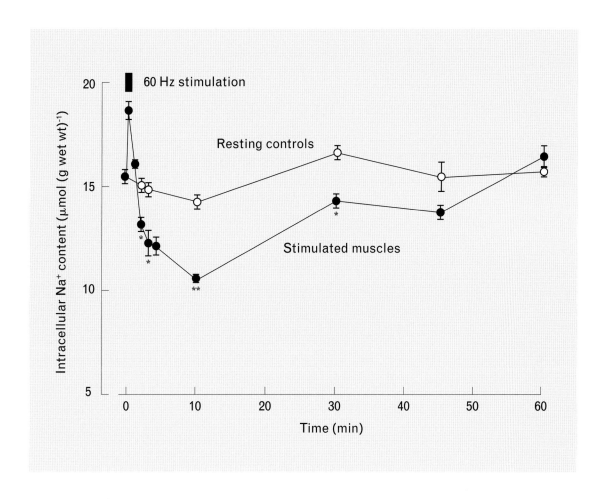

Fig. 17. Time course of changes in intracellular Na⁺ content during or after electrical stimulation in rat soleus. Isolated muscles were mounted for isometric contraction and either stimulated for 30 s at 60 Hz or allowed to rest. At each time point indicated, muscles were immediately transferred to ice-cold Na⁺ free Tris-sucrose buffer, washed 4x15 min, blotted, and taken for determination of Na⁺ content. Open symbols, resting controls; closed symbols stimulated muscles. Each point represents a mean value ± SEM of measurements on 6-27 muscles (Nielsen and Clausen, 1997).

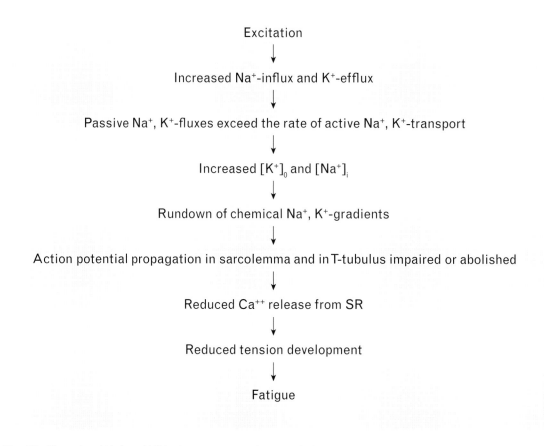

Excitation

↓

Increased Na$^+$-influx and K$^+$-efflux

↓

Passive Na$^+$, K$^+$-fluxes exceed the rate of active Na$^+$, K$^+$-transport

↓

Increased $[K^+]_0$ and $[Na^+]_i$

↓

Rundown of chemical Na$^+$, K$^+$-gradients

↓

Action potential propagation in sarcolemma and in T-tubulus impaired or abolished

↓

Reduced Ca^{++} release from SR

↓

Reduced tension development

↓

Fatigue

Fig. 18. The role of Na$^+$ and K$^+$ in the sequence of events linking excitation with tension development and fatigue in skeletal muscle during high frequency stimulation (Clausen and Nielsen, *Acta Physiol. Scand.,* 1994).

▶▶ **Fig. 19.** Diagram of excitation-induced exchange of Na$^+$ and K$^+$ between ECV and ICV in an intact rat EDL muscle. See Table 2 in Clausen, J. Gen. Physiol, Jan. 14, 2013, for values of Na$^+$ and K$^+$ in the different compartments. Excitation (left) is induced by a rapid influx of Na$^+$ (11.2 µmol/g wet wt.) from ECV into ICV, followed by an equivalent efflux of K$^+$ (11.0 µmol/g wet wt.) from ICV into the ECV. This takes place while the muscles are in air, and during the following washout in ice-cold Na$^+$-free Tris-sucrose, 11 µmol/g wet wt. of K$^+$ is removed from the ECV as can be seen from the loss of 11µmol/g wet wt. while the Na$^+$ taken up into the ICV stays in the ICV, leading to a reduction in total K$^+$ content and a similar increase in total Na$^+$. As shown in the same publication, the Na$^+$, K$^+$ contents are restored by stimulation of the Na$^+$, K$^+$ pumps during the following 600 sec incubation without buffer. When the muscles are incubated in air, these movements of Na$^+$ and K$^+$ may be followed undisturbed of exchange with a surrounding incubation medium.

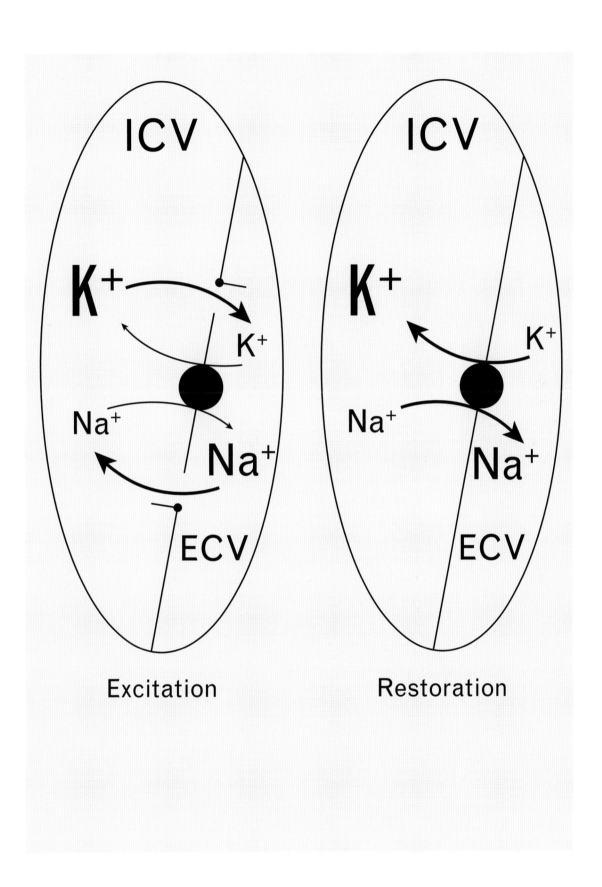

Excitation   Restoration

49

Acute and long-term regulation of Na^+, K^+ pumps in skeletal muscle

As summarized in Fig. 16, the $Na^{+,}$ K^+ pumps are subject to two major and different modes of regulation: a) by acute changes in the rate of active Na^+, K^+ transport (over seconds to minutes) or b) by long-term changes in the content (transport-capacity) of Na^+-K^+ pumps of the muscles (over minutes to days). Excitation causes a graded and often rapid stimulation of the Na^+, K^+ pumps, dependent on the frequency and ranging from increases of a few per cent up to around 2000 per cent, sufficient to activate all the Na^+, K^+ pumps in the muscle (Fig. 8). The precise time-course of stimulation or inhibition of the Na^+, K^+ pumps as well as the Na^+,K^+ pump capacity may be a question of life or death for the intact organism. In isolated rat muscles, the excitation-induced rise in $[Na^+]_i$ has been shown to be followed by an "undershoot" in $[Na^+]_i$ indicating that it reflects a rapid or even lasting increase in the affinity of the Na^+, K^+ pumps for $[Na^+]_i$ causing $[Na^+]_i$ to decrease below the resting level (Fig. 17).

The time-course of the stimulating effect of increased $[Na^+]_i$ has been explored by graded loading of rat or mouse muscles with Na^+ by inhibiting the Na^+,K^+-pumps by preincubation in K^+ free buffer, followed by washout in the cold (Clausen & Kohn, 1977). This leads to progressive accumulation of intracellular Na^+ and ensuing stimulation of the Na^+- K^+ pumps. Isolated muscles may also be loaded with Na^+ using pre-incubation with the Na^+ ionophore A23187 or with the plant alkaloid veratridine, which makes the Na^+ channels stay open.

Since active Na^+ K^+-transport in isolated rat soleus muscle has a Q_{10} of 2.3, the rate of Na^+, K^+-pumping may be about doubled by augmenting the temperature of the incubation medium by 10 °C. Indeed, changing incubation temperature from 20 to 30 °C was found to increase ouabain-suppressible ^{86}Rb uptake in rat soleus by 93 % and to cause repolarization (Pedersen et al. 2003).

Fig. 18 illustrates the time-course of the events following excitation, changes in Na^+ and K^+ fluxes, $[Na^+]_i$ $[K^+]_o$, Na^+, K^+ gradients, action potentials, Ca^{2+} release, reduced tension development and fatigue in isolated rat muscle (Clausen & Nielsen, 1994).

Recent studies of excitation-induced exchange of Na$^+$ and K$^+$ between the extracellular and intracellular water spaces have shown that when rat EDL muscles incubated in KR buffer are transferred to incubation chambers containing only air without KR buffer, and stimulated for 60 s at 60 Hz there is a rapid influx of Na$^+$ of 11 µmol/g wet wt, followed by an equivalent efflux of K$^+$ from the muscle cells into the extracellular water space and subsequent incubation without buffer for 600 s, allowing restoration of K$^+$ into the muscle cells and reextrusion of Na$^+$ from the muscle cells (Clausen, 2013 and Fig. 19 in the present text).

Members of the FXYD family have been identified as tissue-specific regulators of the Na$^+$ K$^+$-ATPase (Fuller et al, 2004). Thus rat cardiac Na$^+$, K$^+$-ATPase was shown to increase 3-fold after 30 min of ischemia compared with aerobic controls. This was shown to be mediated by phospholemman.

In conclusion, isolated intact muscles may be used to quantify the capacity for active Na$^+$ K$^+$-transport as modified by temperature, anoxia, animal species, muscle type, hormones, drugs, training and age.

►► **Fig. 20.** Sequence of events in the action of catecholamines and β$_2$-agonists on active Na$^+$, K$^+$ transport, Na$^+$, K$^+$ distribution, and contractile performance in skeletal muscle. Epinephrine, norepinephrine, or β$_2$-agonists bind to the β$_2$-adrenoceptors, which activate the adenylyl cyclase to stimulate the conversion of ATP to cAMP. This second messenger in turn activates protein kinase A, which is assumed to induce a conformational change in the Na$^+$, K$^+$ ATPase. This leads to increased affinity for intracellular Na$^+$, allowing for an acute increase in Na$^+$ efflux and subsequent maintenance of a steeper Na$^+$ gradient across the sarcolemma. Due to the electrogenic nature of the coupled active Na$^+$, K$^+$ transport, the early increase in Na$^+$ efflux leads to hyperpolarization, increased intracellular K$^+$/Na$^+$ ratio, decreased localized $[K^+]_o$ and hypokalemia. These changes may augment excitability and contractile performance (Modified from Clausen, 2003, *Physiol Rev.*).

►►► **Fig. 21.** Effects of epinephrine and ouabain on ^{22}Na efflux. Isolated soleus muscles were washed and incubated for 60 min in Krebs-Ringer bicarbonate buffer containing ^{22}Na (6.7 µCi/ml) without or with ouabain (10^{-3} M). The washout of isotope was then followed by transferring the muscles through a series of tubes containing unlabeled buffer without or with the additions indicated. The fraction of ^{22}Na released/ min was calculated as previously reported (Clausen and Kohn, 1977), and each curve represents the mean of 3-6 observations with bars denoting S.E. of the mean. Epinephrine (6×10^{-6}M) was present in the efflux medium during the interval indicated by the horizontal bar and the filled symbols (Clausen and Flatman, *J. Physiol.* 1977).

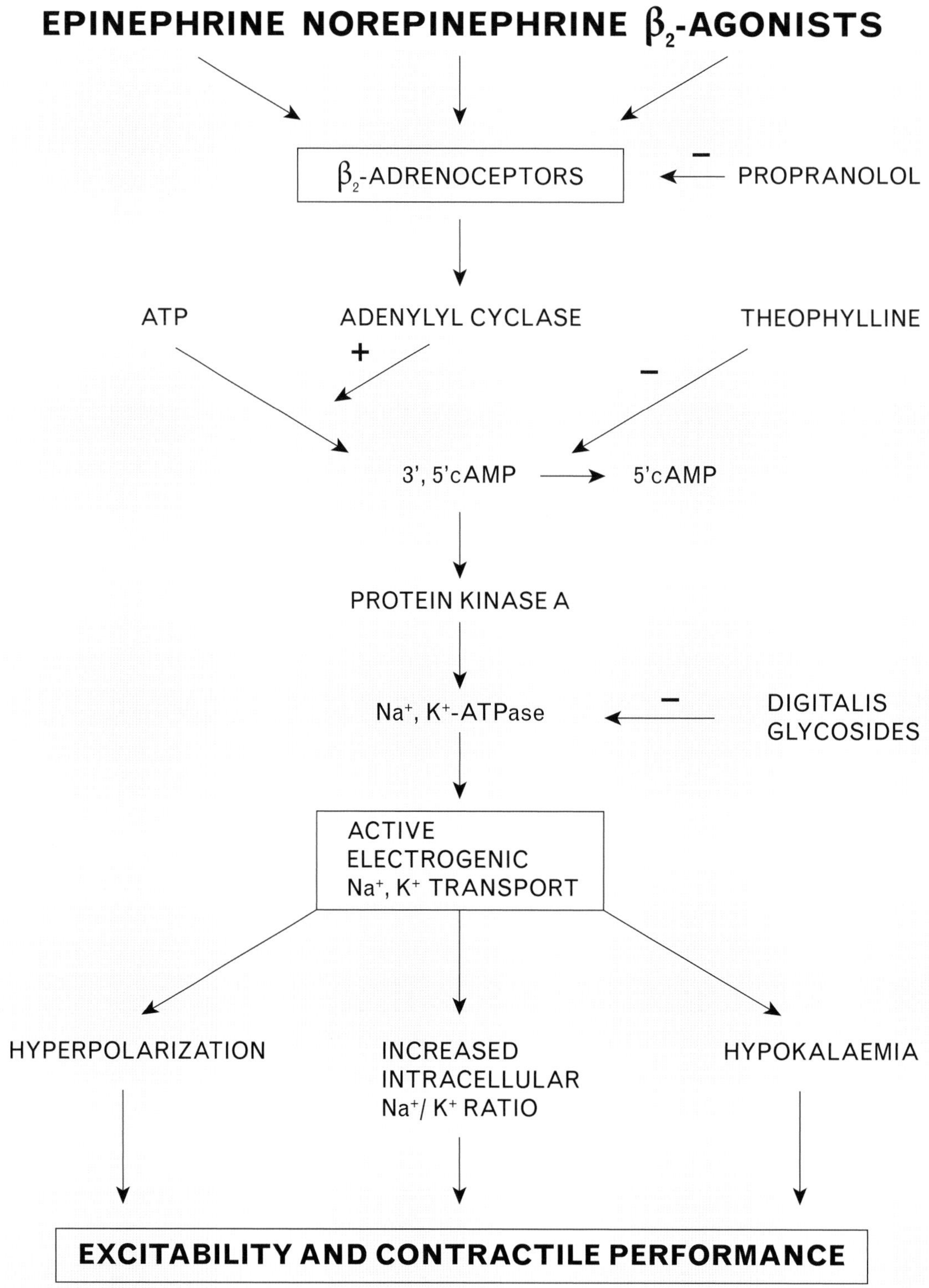

EPINEPHRINE NOREPINEPHRINE β_2-AGONISTS

β_2-ADRENOCEPTORS $\xleftarrow{\quad-\quad}$ PROPRANOLOL

ATP   ADENYLYL CYCLASE   THEOPHYLLINE
                    $+$ $-$

3', 5'cAMP $\longrightarrow$ 5'cAMP

PROTEIN KINASE A

Na^+, K^+-ATPase $\xleftarrow{\quad-\quad}$ DIGITALIS GLYCOSIDES

ACTIVE
ELECTROGENIC
Na^+, K^+ TRANSPORT

HYPERPOLARIZATION   INCREASED
                           INTRACELLULAR
                           Na^+/ K^+ RATIO   HYPOKALAEMIA

EXCITABILITY AND CONTRACTILE PERFORMANCE

Fig. 20

52

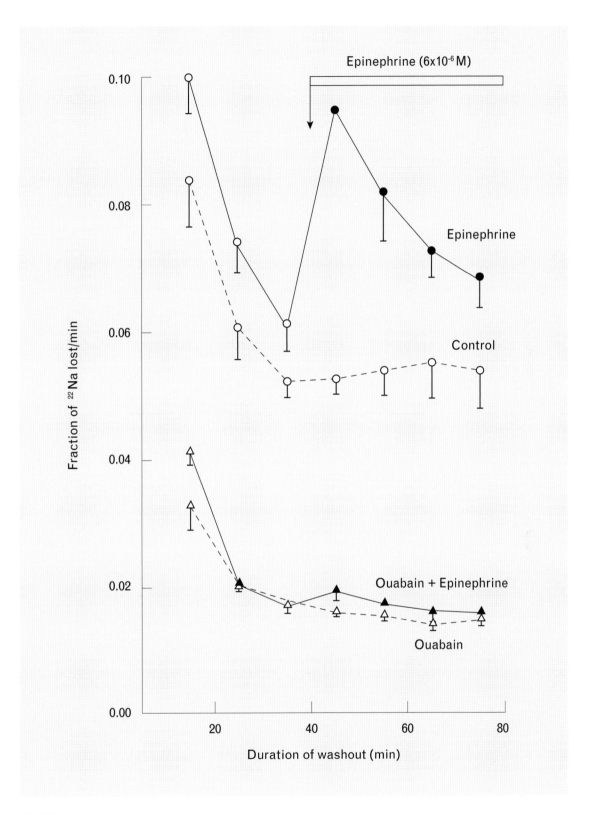

Fig. 21

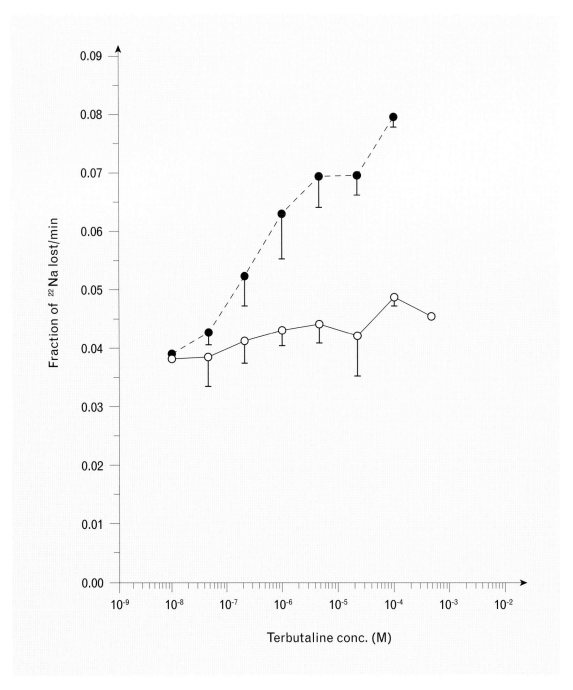

Fig. 22a

Fig. 22. Effects of terbutaline (a) and epinephrine (b) on ^{22}Na-efflux from soleus muscles from guinea-pigs pretreated with terbutaline (——) or control animals (--------). The muscles were loaded for 90 min by incubation in 3 ml Krebs-Ringer bicarbonate buffer containing 3-5 μC/ml of ^{22}Na. The muscles were then washed out in a series of tubes containing non-radioactive buffer; 40 min after the start of

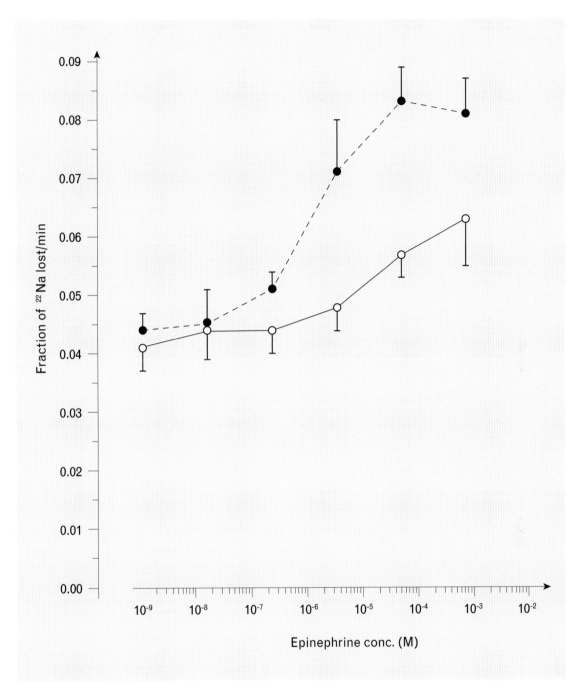

Fig. 22b

washout, terbutaline or epinephrine were added to the efflux medium at the concentrations indicated. The mean fractional loss of ^{22}Na as determined from 10 to 20 min after the start of exposure to the drugs is shown. n = 2-4 in (a) and 4-5 in (b); vertical lines indicate s. e. of the mean. (Buur et al., *Br. J. Pharm*. 76:313-317, 1982)

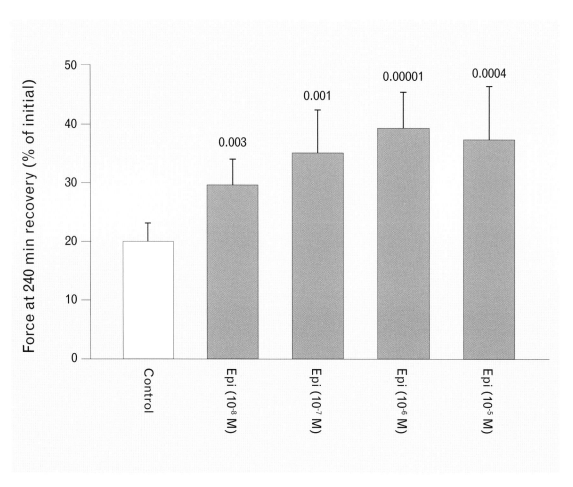

Fig. 23. Dose-response of the action of epinephrine ($10^{-8} - 10^{-5}$ M) on force recovery after 30-min fatiguing stimulation. EDL muscles were fatigued by being stimulated intermittently (10 s on at 40 Hz, 1 ms pulses of 10 V; 30 s off). Force after 240 min recovery is given as % of initial force. Epinephrine ($10^{-8} - 10^{-5}$ M) was added after 130 min recovery. To prevent metabolic breakdown of epinephrine ascorbic acid (1 mM) and EDTA (0.05 mM) were added to the buffer. Mean values are given with bars denoting SD. Number of muscles is 4-10. The significance of the difference between the groups was ascertained using a one-way ANOVA (P<0.001), and subsequently, the difference between the control group and any of the pre- treated groups was tested using a Student's t-test. P values are given in the figure. (Mikkelsen et al. *Am. J. Physiol.*, 2006)

Catecholamines and caffeine

Already long ago, catecholamines were shown to favour net Na^+ extrusion and to induce hyperpolarization in isolated rat soleus muscle (Dockry et al., 1966). These effects were mediated via β-adrenoceptors and appeared to be general by being also seen in rat diaphragm, frog sartorius (Hays et al.,1974), avian muscle (Somlyo & Somlyo, 1969), guinea pig soleus (Buur et al., 1982) and EDL muscles (Tashiro, 1973). Since they were suppressed by ouabain, they were attributed to stimulation of the Na^+, K^+ pumps. A detailed analysis showed that epinephrine and norepinephrine within minutes accelerate ^{22}Na efflux and ^{42}K influx. This was accompanied by rapid hyperpolarization (5-8 mV), increased $[K^+]_i$ (by a few per cent) and decreased $[Na^+]_i$ (by up to 67 %) (Clausen & Flatman, 1977; Clausen & Flatman, 1987). Fig. 20 summarizes the time-course of most of these effects of epinephrine on isolated rat muscle. Fig. 21 shows the effects of a supramaximal concentration of epinephrine on ^{22}Na efflux from rat soleus muscles in the absence or the presence of a supramaximal inhibitory concentration (10^{-3} M) of ouabain. Ouabain alone reduces the fractional loss of ^{22}Na by around 70 % and almost abolishes the stimulating effect of epinephrine on the Na^+, K^+ pumps.

Fig. 22 illustrates the stimulating effects of graded doses of terbutaline or epinephrine ($10^{-7} - 10^{-4}$) on the efflux of ^{22}Na from soleus muscles obtained from guinea pigs pretreated with terbutaline (full lines) or control animals (dashed lines) (Buur et al., 1982). These effects were both clearly reduced by 5 days of oral pretreatment with terbutaline, indicating desensitization towards terbutaline and epinephrine. In contrast, in the same study the stimulating effect of insulin on ^{22}Na-efflux was not reduced by pretreatment with terbutaline indicating the selectivity of the desensitization. The desensitization towards the effects of $β_2$-agonists on Na^+, K^+-pumps is of fundamental clinical significance, considering the fact that millions of asthma patients are exposed to these pharmaceuticals every day. For instance it implies that the stimulating effects of $β_2$-agonists on the Na^+, K^+ pumps is reduced, impairing the clearance of $(K^+)_o$.

Also in vivo, epinephrine, salbutamol and isoprenaline induced a hyperpolarization in skeletal muscle cells of up to 15 mV (Flatman & Clausen, 1979; Clausen & Flatman, 1980; Iannaconne et al., 1989). All these effects were mediated by beta$_2$ adrenoceptors (Wang & Clausen, 1976; Clausen & Flatman, 1980), were blocked by ouabain, potentiated by theophylline (a phosphodiesterase inhibitor that inhibits the breakdown of cAMP) and mimicked by cAMP or dibutyryl cAMP (Clausen & Flatman, 1977). Epinephrine and the beta$_2$-agonist terbutaline both increased the content of cAMP in isolated guinea pig soleus). The chain of events from the binding of catecholamines to beta$_2$-adrenoceptors to the changes in Na^+, K^+-fluxes, Na^+, K^+-distribution, membrane potential and muscle contractility is summarized in Fig. 6 in Clausen, 2003 and in Fig. 20 in the present text.

The effect of cAMP is mediated by activation of protein kinase A, which in turn activates the Na^+, K^+ pumps (Shattock, 2009). This scheme seems to be general. The central general role of cAMP in mediating Na^+,K^+ pump activation by other hormones is exemplified by the observations that peptide hormones (calcitonins, CGRP and amylin) that cause increase in cAMP in skeletal muscle also stimulate the Na^+, K^+ pumps (see section 13).

Epinephrine also increases the rate of [^{3}H]ouabain binding, a specific indication that the Na^+, K^+ pumps are stimulated (Clausen & Hansen, 1977). The effects of epinephrine on active Na^+, K^+ transport (6×10^{-9} M) and membrane potential (6×10^{-10} M) of isolated rat soleus are seen at concentrations down to the physiological levels reached in plasma during exercise in human subjects (2.4×10^{-9} M; Galbo et al., 1975) and during running in the rat (6×10^{-9} M; Richter et al., 1981).

Epinephrine (10^{-6} M) induces 10 mV hyperpolarization and a decrease in the activity of intracellular Na^+ in isolated rat soleus and human intercostal muscle fibers, effects that were blocked by ouabain (Ballanyi & Grafe, 1988). Also in vivo, epinephrine stimulated the uptake of K^+ in dog gracilis muscle via a β-adrenoceptor-mediated mechanism (Powell & Skinner, 1966), and in the human forearm a β$_2$ agonist stimulated the net uptake of K^+ in the skeletal muscles (Ford et al., 1995). Intravenous injection of epinephrine or the beta$_2$-adrenoceptor agonist salbutamol both induce hyperpolarization of rat soleus as measured in vivo (Flatman & Clausen, 1979; Clausen & Flatman, 1980), and this effect was offset by the selective β$_2$ adrenoceptor blocker propranolol (Clausen & Flatman, 1980; McCarter et al., 2001). In rat soleus salbutamol increased the uptake of ^{86}Rb by 83 % in vivo (Cartana & Stock, 1995).

In isolated rat soleus, the hyperpolarizing effect of supramaximal concentration of epinephrine (10^{-5} M) was larger and earlier in onset than that of a maximum concentration of insulin (100 mU/ml) (Clausen & Flatman, 1980). In isolated rat soleus the β-agonist isoproterenol (10^{-6} M) induced 9 mV hyperpolarization (Li and Sperelakis, 1993). This effect was mimicked by the adenylate cyclase activator forskolin (10^{-5} M) or 8-bromo-cAMP (1 mM) and blocked by ouabain. Insulin (4 mU/ml) also induced hyperpolarization, and this effect was additive to that of isoproterenol, indicating that these two agents act via different mechanisms. The hyperpolarizing effect of insulin is seen at concentrations down to those found in human blood plasma (10μU/ml) (Li & Sperelakis, 1993).

Studies on isolated rat skeletal muscle myoballs showed that isoproterenol rapidly stimulates ouabain-suppressible transmembrane current by 50 %. This effect was mimicked by 8-bromo-cAMP and attributed to a cAMP-mediated stimulation of the Na^+, K^+ pumps (Li & Sperelakis, 1993). These observations on single cells are important because they provide strong evidence that the hyperpolarization is not entirely secondary to localized decreases in interstitial $[K^+]_o$. In rat soleus, Ba^{2+} (which blocks the K^+ channels) does not suppress the hyperpolarizing effect of salbutamol, indicating that this is not mediated by an action on K^+-channels (Cox et al. 1978). The effects of salbutamol and CGRP on ^{42}K uptake and $[Na^+]_i$ in soleus were both independent of Ba^{2+}-sensitive and ATP-sensitive K^+ channels (Clausen & Overgaard, 2000). Taken together, the stimulating effects of catecholamines on the Na^+, K^+ pumps in skeletal muscle seem to account for the rapid hyperpolarization of muscle cells.

During intense exercise, plasma catecholamines increase markedly (Galbo et al., 1975; Brooks et al. 1990), reflecting rapid stimulation of secretion from the adrenal medulla and release of norepinephrine from the sympathetic nerve endings. These endogenous catecholamines are likely to counterbalance hyperkalemia by stimulating the uptake of K^+ via the Na^+, K^+ pumps in resting or relatively inactive muscles (Clausen & Flatman, 1977; Everts et al., 1988). When exposed to elevated $[K^+]_o$ (11 mM), the muscle cells in rat soleus were depolarized and contractile force reduced to 25 % of control force measured at 4 mM K^+.

As shown in Fig. 23, rat EDL muscles exposed to 30 min of fatiguing stimulation leading to a considerable loss of force, cell leakage, and depolarization, stimulation of the Na^+, K^+ pumps with epinephrine (10^{-8} – 10^{-5} M) induces highly significant force recovery, possibly due to the electrogenic action of the Na^+, K^+-pumps (Mikkelsen et al., 2006). This mechanism may be important for restoration of muscle function after intense exercise.

In nine healthy non-coffee drinkers, oral ingestion of caffeine (250 mg, corresponding to 2-3 cupfuls of coffee) was found to increase plasma epinephrine by 207 % (Robertson et al., 1978). Intoxication with caffeine was shown to induce massive release of endogenous catecholamines and hypokalemia with plasma K^+ down to 2.2 mM (Benowitz et al., 1982). Another patient who received an overdose of caffeine unexpectedly contained in a slimming agent showed a plasma concentration of caffeine of 110 mg/L and a plasma-K^+ of 1.6 mM (de Wijkerslooth et al., 2008).

In conclusion, in view of the widespread intake of coffee and β_2-agonists, these effects of caffeine and β_2-agonists are important, in particular for subjects treated with K^+ loosing diuretics, on inadequate dietary K^+-intake or with diarrhea. As an increase in endogenous or therapeutic plasma β_2-agonists stimulate the Na^+, K^+-pumps, high doses of caffeine are likely to worsen the risks of hypokalemia.

Peptide hormones stimulating the Na^+, K^+ pumps

Soon after insulin was purified in the 1920s and shown to induce hypoglycemia it was also found to induce hypokalemia. This was the start of investigation of the effects of insulin and other peptide hormones on the transport and distribution of Na^+ and K^+ in skeletal muscle. Insulin was shown to decrease $[Na^+]_i$ in rat diaphragm (Creese, 1968). Similar observations obtained in frog sartorius (Moore, 1973) and rat soleus (Clausen & Kohn, 1977), led to the proposal that this hormone acts by increasing the affinity of the Na^+, K^+ pumps for intracellular Na^+ (Creese, 1968). This suggestion was documented by ^{22}Na flux studies on frog sartorius showing that insulin induced a clear-cut leftward shift of the relationship between Na^+ efflux and the logarithm of $[Na^+]_i$ concentration (Kitasato et al., 1980). In rat adipocytes, insulin stimulated the Na^+, K^+ pumps without any change in V_{max}, but by augmenting the affinity for intracellular Na^+ (Lytton, 1985). In keeping with this, quantification of α_1- and α_2-subunit isoforms of Na^+, K^+-ATPase with immuno-gold labeling showed that insulin produced no increase in the Na^+, K^+- ATPase content of the plasma membrane, even under conditions where the hormone induced a sevenfold increase in the content of the glucose transport system Glut4 (Voldstedlund et al., 1993). This is in keeping with the observation that in rat adipocytes, insulin produced no change in the total content of [^{3}H]ouabain binding sites (Clausen & Hansen, 1977; Resh et al., 1980). In cultured human fibroblasts that were shown to contain only the α_1-subunit isoform of Na^+, K^+-ATPase, insulin clearly increased ^{86}Rb uptake but caused no change in the content of [^{3}H]ouabain binding sites (Longo et al. 2001). In rabbit cardiac myocytes, insulin stimulates Na^+, K^+ pump-mediated current by 70 % (Hansen et al., 2000). This effect was abolished when $[Na]_i$ was increased to 80 mM, leading to saturation of the Na^+, K^+ pumps. This indicates that in cardiac myocytes, the effect of insulin is not due to translocation of the Na^+, K^+ pumps to the plasma membrane but rather reflects an increase in the affinity of the Na^+, K^+ pumps for $[Na^+]_i$. Likewise, in isolated rat proximal convoluted kidney tubules, the stimulating effect of insulin on ouabain-suppressible ^{86}Rb uptake and Na^+, K^+-ATPase activity disappeared when $[Na^+]_i$ was increased to saturating levels. Because insulin decreased the apparent dissociation constant for Na^+ from 47 to 28 mM, it was concluded that the hormone acted by increasing the sensitivity of the Na^+, K^+-ATPase to Na^+ and not by increasing V_{max} (Feraille et al.,1994).

Insulin stimulates the uptake of ^{22}Na in rat soleus (Clausen & Kohn, 1977; Clausen & Flatman, 1987). This effect was not reduced by amiloride or bumetanide, indicating that it is not the result of stimulation of the amiloride-sensitive Na^+/H^+ antiporter or the bumetanide-sensitive NaCl cotransport system found in rat skeletal muscle (Dørup & Clausen, 1994). Also in human fibroblasts, insulin stimulated Na^+ uptake by a bumetanide-resistant mechanism (Longo et al. 2001).

In rat soleus incubated in the presence of amiloride, insulin still produced a clear-cut stimulation of ^{86}Rb uptake as well as ^{22}Na efflux (Clausen & Flatman, 1987; Hsu & Guidotti, 1991). Likewise, in primary cultures of rat skeletal muscle, neither the stimulating effect of insulin on ^{86}Rb uptake nor its hyperpolarizing effect were reduced by amiloride (Sampson et al., 1994). The effect of insulin on the uptake of ^{42}K was blocked by ouabain (Clausen & Flatman, 1977; Clausen & Kohn, 1977). In the human forearm, the stimulating effect of an intra-arterial infusion of insulin on the net uptake of K^+ was abolished by subsequent infusion of ouabain (Ferrannini et al., 1988). In contrast, the effect of insulin on glucose uptake was preserved. This is in keeping with studies on isolated rat diaphragm (Clausen, 1966) and soleus (Clausen & Flatman, 1987), showing that the transport of glucose is not coupled to active Na^+, K^+ transport or its response to insulin. In rat soleus, insulin stimulated the uptake of ^{86}Rb and decreased $[Na]_i$, effects that were both blocked by ouabain (Weil et al., 1991), but not by amiloride or the Na^+ ionophore monensin. Insulin also increased the rate of $[^3H]$ouabain binding to rat soleus, a rather specific indication that the rate of active Na^+, K^+ transport is accelerated. In vivo recordings of the resting membrane potential of the rat soleus showed that insulin injected intravenously produced a hyperpolarization of 15 mV, which was additive to the hyperpolarizing effect of salbutamol (Flatman & Clausen, 1979). Also endogenous insulin released by injection of glucose produced a significant hyperpolarization of 5.5 mV. A similar effect of intravenous insulin on the in vivo recorded membrane potential of soleus was observed (Iannaconne et al., 1989). This effect was additive to that of isoproterenol and blocked by ouabain. In soleus of diabetic rats, insulin produced only 70 % of the hyperpolarization seen in normal muscles. In primary cultures of rat skeletal muscle cells, both insulin and phorbol esters induced stimulation of ^{86}Rb uptake and hyperpolarization (Sampson et al., 1994). These effects were not additional, but suppressed by staurosporine or by prior downregulation of protein kinase C. This indicates that insulin stimulates the Na^+, K^+ pump activity via activation of protein kinase C. In keeping with this, phorbol ester induced a ouabain-suppressible hyperpolarization in isolated rat soleus (Li & Sperelakis, 1994). In the rat skeletal muscle cell line L6, incubation for 5 min with insulin (100 nM) produced a significant increase in the activity of Na^+, K^+-ATPase measured in plasma membranes (Ragolia et al., 1997). When taken together, these observations from several different laboratories strongly indicate that the classical hyperpolarizing action on skeletal muscle discovered by (Zierler & Rabinowitz, 1964) is the result of stimulation

61

of the electrogenic Na^+, K^- pump. Stimulating effects could be detected in the insulin concentration range 0.1-100 nM and reached a maximum at 100 % after 15 min. The effect was associated with dephosphorylation of the α_1-subunit isoform, mediated by a serine/threonine protein phosphatase. In the rat, however, intravenous administration of insulin produced no change in the content of the α_1- subunit isoform in the plasma membrane and t-tubular fractions prepared from the hindlimb muscles, but an 80 % increase in the content of the glucose transporter Glut4 (Dombrowski et al., 1996).

The physiological significance of the effects of insulin on the Na^+, K^+ pumps in skeletal muscle has been documented by the following:

○ Down to physiological plasma concentration levels (10 μU/ml), insulin stimulates the net uptake of K^+ in the human forearm (Zierler & Rabinowitz, 1964).

○ During oral or intravenous glucose loads, the endogenous insulin release is sufficient to promote the clearance of K^+ from plasma (Natali et al., 1993).

○ Acute inhibition of insulin secretion by the infusion of somatostatin causes a slow increase in plasma K^+ of 0.5 mM (DeFronzo et al., 1978), indicating that basal insulin secretion/plasma levels are required for the maintenance of normal plasma K^+.

○ Hyperkalemia has repeatedly been shown to increase plasma insulin (Cox et al.1978; Clausen et al. 1980; Clausen, 1986).

For these reasons, it is likely that insulin via its stimulating effect on the affinity of the Na^+, K^+ pump for Na^+ contributes significantly to K^+ homeostasis in skeletal muscle as well as in the entire organism.

In isolated frog sartorius muscle, insulin was reported to increase the binding of [3H]ouabain, which was taken to indicate a mobilization of Na^+, K^+ pumps from an intracellular pool to the membrane surface (Erlij & Grinstein, 1976). However, the measurements of [3H]ouabain binding were performed before it could be ascertained that all Na^+, K^+ pumps were occupied. Later studies indicated that the observation was likely to be the result of insulin-induced stimulation of the rate of [3H]ouabain binding. Thus, it could be shown that when binding equilibrium had been reached, insulin caused no significant change in [3H]ouabain binding (Clausen and Hansen, 1977; Dørup & Clausen, 1995). More recently, it was found that in control subjects and type 2 diabetics, infusion of insulin with glucose caused no significant change in the content of [3H]ouabain binding sites in biopsies obtained from the vastus lateralis muscle (Djurhuus et al., 2001). In isolated intact rat soleus, pretreatment with insulin produced no increase in the binding of [3H]ouabain to muscle specimens prepared after the incubation (McKenna et al., 2003).

In isolated frog skeletal muscles, preincubation with insulin (100 mU/ml) for 30 min induced an 85 % increase in the content of Na^+, K^+-ATPase enzyme activity in a plasma membrane fraction subsequently prepared from the same muscles (Kanbe & Kitasato, 1986; Omatso-Kanbe & Kitasato, 1987). After preincubation of the muscles with ouabain for 60 min, the plasma membrane fraction contained no Na^+, K^+-ATPase. When the incubation with ouabain was followed by 60 min incubation with insulin, the plasma membrane fraction contained the same Na^+, K^+-ATPase activity as in the untreated control muscles.

This reappearance of Na$^+$ K$^+$-ATPase was taken to indicate an insulin-induced translocation of Na$^+$, K$^+$ pumps from an intracellular pool to the plasma membrane. This cannot be ascertained, however, because it was not tested whether muscles incubated without insulin showed a similar reappearance of Na$^+$, K$^+$-ATPase, due to washout of ouabain during the 60-min incubation. In frog skeletal muscle, preincubation with insulin for 60 min produced a 95 % increase in the Na$^+$, K$^+$-ATPase activity of the plasma membrane and a corresponding increase in the [^{3}H]ouabain binding capacity of the plasma membrane (Omatsu-Kanbe & Kitasato, 1990). Concomitantly, there was a decrease in the content of Na$^+$, K$^+$-ATPase and [^{3}H]ouabain binding sites in an intracellular membrane fraction. The recovery of plasma membrane was only 4 % but not changed by insulin treatment. The muscles used and the conditions of incubation were not defined. In hindlimb muscles prepared from insulin-pretreated rats, the content of the α_2 subunit isoform of the plasma membrane fraction was doubled, and in an intracellular membrane fraction there was a similar decrease (Hundal et al. 1992). The α_1-isoform was almost entirely confined to the plasma membrane fraction and unaffected by insulin. It was proposed that the α_1-isoform mediated the basal active Na$^+$,K$^+$ transport, whereas the α_2-isoform was the regulated isoform.

Insulin-induced translocation of the α_2 subunit isoform Na$^+$-K$^+$ pumps is only seen in type I fibers, and not in muscles containing predominantly type II fibres (Lavoie et al., 1996). This is surprising, in view of the fact that in muscles containing predominantly type II fibers, insulin stimulates active Na$^+$, K$^+$ transport (Erlij & Grinstein, 1976; Moore, 1973).

By immuno-electron microscopy, the effect of insulin on the content of α_2-isoform subunit was assessed in rat skeletal muscle. 30 min after an intravenous injection of insulin, the content of immunolabeled α_2-subunit per linear micron of plasma membranes had increased 1.5- to 3.7-fold (Marette et al, 1993). Unfortunately, these data do not allow accurate quantification of the content of Na$^+$- K$^+$ pumps per square micron of plasma membrane or per gram of muscle. In the isolated split rat soleus preparation, insulin induced a 39 % increase in the abundance of the α_2-subunit isoform in plasma membranes prepared by differential centrifugation (Chibalin et al., 2001).

In normal mice, injection of insulin caused no significant change in the content of the α_2-subunit isoform of the Na$^+$, K$^+$-ATPase detected in immunoblots of the plasma membranes prepared from hindlimb skeletal muscle. However, insulin induced a small but significant increase in the β_1-subunit of normal mice and a much larger increase in muscles obtained from transgenic mice overexpressing the Glut4 glucose transporter (Ramlal et al., 1996).

In conclusion, the Na$^+$, K$^+$ pumps in skeletal muscle are stimulated within minutes by insulin over a range of concentrations down to the physiological level. This leads to hyperpolarization and redistribution of Na$^+$ and K$^+$ contents and is important for whole body K$^+$ homeostasis and contractile performance. The stimulation of the Na$^+$, K$^+$ pumps depends on protein kinase C and seems to reflect an increase in the affinity of the enzyme for intracellular Na$^+$. There is some evidence that insulin induces a recruitment of Na$^+$, K$^+$ pumps by translocation from an undefined intracellular pool to the sarcolemma.

Insulin-like growth Factor-I (IGF-I)

IGF-I is a polypeptide with structural homology to proinsulin. It interacts with the type I IGF receptor and to a minor extent also with the structurally similar insulin receptor. In lambs, the infusion of recombinant human IGF-I and insulin in doses resulting in the same relative drop in plasma glucose caused almost the same reductions (15-18 %) in plasma K^+ (Douglas et al., 1991). Also in normal human subjects, IGF-I induced hypokalemia (Giordano & DeFronzo, 1995). The mechanism of this hypokalemic effect was explored in experiments with isolated rat soleus and EDL muscles (Dørup & Clausen, 1995). IGF-I stimulated ^{22}Na efflux as well as ^{42}K and ^{86}Rb influx, leading to increased $[K^+]_i$ and decreased $[Na^+]_i$ (21-55 %). All these effects, except that on ^{22}Na efflux,

were blocked by ouabain (10^{-3} M). IGF-I decreased $[Na^+]_i$ and increased intracellular K^+ in EDL muscle to the same extent as in soleus. IGF-I increased the rate of $[^3H]$ouabain binding but had no detectable effect on the total binding capacity for $[^3H]$ouabain. The effects of IGF-I and insulin on ^{86}Rb uptake and $[Na^+]_i$ were not additive. In contrast, the effects of IGF-I and epinephrine on the same parameters were additive. This indicates that IGF-I and insulin stimulate the Na^+, K^+ pumps via the same pathway, which is different from that activated by epinephrine.

In conclusion, these observations indicate that IGF-I may participate in the regulation of Na^+, K^+ homeostasis of skeletal muscle, but more detailed in vivo studies are required to substantiate this possibility.

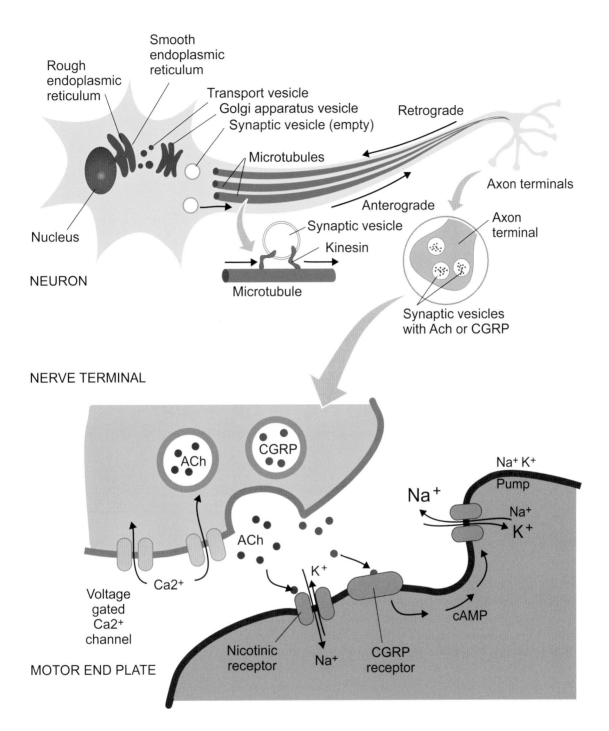

Fig. 24. CGRP is situated in neurons and in rat sciatic nerves. It is transported from the nerve cells in both motor and sensory nerves. From these peripheral depots, CGRP may be released during electrical stimulation, exposure to high K^+ or capsaicin. In the motor end-plate, CGRP is bound to its receptor, stimulating adenylate cyclase, mediating the stimulatory effect of catecholamines on the Na^+, K^+ pumps. *(Clausen, Andersen & Flatman, Am. J. Physiol., 1993).*

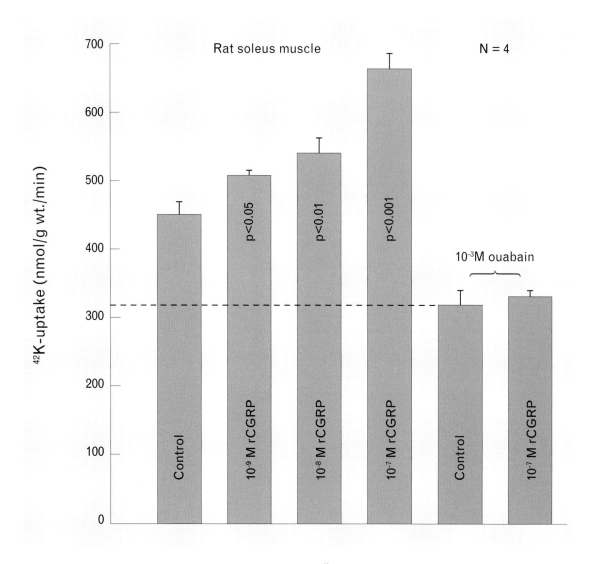

Fig. 25. Effect of varying concentrations of rat CGRP on ^{42}K uptake in rat soleus muscle during 10 min of incubation. Groups of 4 muscles were preincubated for 15 min in the absence or presence of ouabain (10^{-3} M) and then transferred to polyethylene counting vials containing 4 ml Krebs-Ringer bicarbonate buffer with 1% bovine serum albumin, ^{42}K (0.1 µCi/ml) and the indicated additions. After gassing with a mixture or 95% O_2 and 5% CO_2, vials were capped and incubated for 10 min in a metabolic shaker. After incubation, muscles were washed for 4x 15 min in ice-cold Na^+-free Tris-sucrose buffer. Amount of ^{42}K taken up and retained during the cold wash was determined. Each column represents the mean of 4 observations, with bars denoting SEM. (Andersen and Clausen, *Am. J. Physiol.*, 1993).

CGRP and other calcitonins

CGRP is a 37-amino acid peptide present in a variety of central and peripheral neurons. As shown in Fig. 24, CGRP is situated in neurons and in rat sciatic nerves, CGRP is transported in somatofugal direction (from the nerve cells into the nerve fibres) in both motor and sensory nerves and accumulates in nerve terminals in the muscle (Kashihara et al., 1989). From these peripheral depots, CGRP may be released during electric stimulation, exposure to high $[K^+]_o$, or specific agents such as capsaicin. CGRP binds to receptors on sarcolemma and stimulates adenylate cyclase activity leading to a rise in intracellular cAMP (Takami et al., 1986). Because this second messenger mediates the stimulatory effects of catecholamines on the Na^+, K^+- pumps (Clausen & Flatman, 1977), it was anticipated that CGRP would exert a similar action. In rat soleus, rat CGRP stimulated ^{22}Na efflux and the influx of ^{42}K and ^{86}Rb, leading to increased $[K^+]i$, decreased $[Na^+]_i$, and hyperpolarization. As shown in Fig. 25 the stimulating effect of rat CGRP on ^{42}K uptake is significant, detectable down to a concentration of 10^{-9} M and completely blocked by 10^{-3} M ouabain, indicating that

it reflects an acute stimulation of active electrogenic Na^+, K^+ transport (Andersen & Clausen, 1993).

Similar effects were exerted by human CGRP as well as rat and salmon calcitonin. Because the effects of rat CGRP and epinephrine on $[Na^+]_i$ and ^{86}Rb uptake were not additive, they may be exerted via the same messenger (cAMP). In contrast, the effects of rat CGRP and insulin are additive. Capsaicin, which releases CGRP from the sensory nerve endings, induced a 51 % decrease in the CGRP content of rat soleus (Nielsen et al., 1998)(157). The CGRP released is available for action on the muscle cells, and capsaicin elicited effects similar to those of added CGRP on ^{22}Na extrusion, $[Na^+]_i$ and ^{86}Rb uptake (Andersen & Clausen, 1993). These effects were not impaired by propranolol, indicating that they were not the outcome of release of norepinephrine from the sympathetic nerve endings. Moreover, ruthenium red, which inhibits the effect of capsaicin on the release of CGRP, blocked the effects of capsaicin on ^{86}Rb uptake and $[Na^+]_i$. The capsaicin-induced release of CGRP in rat soleus amounts to 4 pmol/ml extracellular fluid, corresponding to a local CGRP concentration of 4×10^{-9} M, sufficient to elicit the above-mentioned stimulation of the Na^+- K^+ pumps (Andersen & Clausen, 1993). This implies that the endogenous pools of CGRP in the nerve endings are adequate to influence Na^+, K^+ homeostasis in skeletal muscle. Exogenously added CGRP (20-200 nM) or endogenously released CGRP also stimulated the short-circuit current in isolated frog skin, an effect that was partially suppressed by an adenylate cyclase inhibitor and therefore proposed to be mediated by cAMP (Castronuovo et al., 1996; Lippe et al., 1994).

The physiological significance of the stimulating effect of CGRP on the Na^+, K^+ pumps is still not completely understood, but it seems most relevant that a localized rise in $[K^+]_o$ induces a release of CGRP, which, in turn, might favor the local clearance of K^+ (Sakaguchi et al., 1991; Santicioli et al., 1992). CGRP is also a trophic factor generated in specific motoneurons. It is interesting that following unaccustomed downhill running (the so-called "fell-running" practiced by the inhabitants on the steep mountain sides in Lake District, UK), the number of CGRP-positive neurons supplying the hindlimb extensors of adult Wistar rats at work increased threefold and the number of CGRP motor end plates increased even more (Homonko & Theriault, 2000). This would allow for more intense submaximal stimulation of the Na^+, K^+ pumps during subsequent bouts of exercise and might thus constitute part of the training adaptation. The excitation-induced release of CGRP from isolated muscles would suggest that during exercise, plasma CGRP increases. Although intense exercise produced no change (Brooks et al., 1990), steady-state exercise gave 50 % increase in plasma CGRP (Onuoha et al., 1998).

In conclusion, there is good experimental basis for the idea that CGRP participates in the regulation of Na^+, K^+ pump activity in skeletal muscle, possibly as a local humoral mediator of nervous control of the Na^+, K^+ pump during exercise and adaptation to localized increases in $[K^+]_o$ or training. In mutant mice with a similar genetic anomaly as that seen in human subjects with hyperkalemic periodic paralysis contractility of isolated soleus muscle was markedly reduced compared to soleus from wild type mice, reflecting impaired excitability (see Fig. 33). This weakness could be completely restored by incubation with CGRP (Clausen et al., 2011) (see sections 23 and 24).

Amylin, related peptides, and other stimuli for the Na^+, K^+-pumps

Amylin is a 37-amino acid peptide present in the pancreatic beta-cells and is cosecreted with insulin into the bloodstream. Its amino acid sequence is 46 % identical to that of CGRP, and it has been shown to mimic several of the actions of CGRP, among others an increase in cAMP in skeletal muscle. This would be expected to mediate stimulation of the Na^+, K^+ pumps (see Fig. 20) and as shown for CGRP, in rat soleus, amylin induces an increase in ^{22}Na efflux and ^{86}Rb influx, leading to a 50 % decrease in $[Na^+]_i$ and a minor increase in $[K^+]_i$ (Clausen, 1996a+b; Clausen, 2000). All these effects were blocked by ouabain, and similar actions were exerted on rat EDL (James et al., 1999). In contrast, neither the structurally similar peptides islet amyloid peptide (IAPP) nor adrenomedullin produced any detectable change in Na^+ or K^+ uptake in rat soleus (Clausen, 2000). Like insulin, CGRP, and epinephrine, amylin also increased the rate of $[^3H]$ouabain binding without causing any change in the total content of $[^3H]$ouabain binding sites in rat soleus and EDL (Clausen, 2000). The effects of supramaximal concentrations of insulin and amylin on ^{86}Rb uptake in rat soleus are additive also on tetanic and twitch contractions at 12.5 mM K^+, indicating that they are exerted via two separate pathways. It is interesting that the two hormones are secreted from the same cells (β) in pancreas. There is good evidence that amylin stimulates the Na^+, K^+ pumps both in type I (in soleus) and type II (in EDL) fibres. This may account for the significant hypokalemic action of the hormone in human subjects (a drop of 16 % in plasma K^+ in 1 h) when injected in large doses (Vine et al. 1998).

In conclusion, the physiological significance of the amylin-induced Na^+, K^+ pump stimulation in skeletal muscle is not yet known, but it represents another example of a hormonal Na^+, K^+ pump activation in skeletal muscle mediated by the second messenger cAMP.

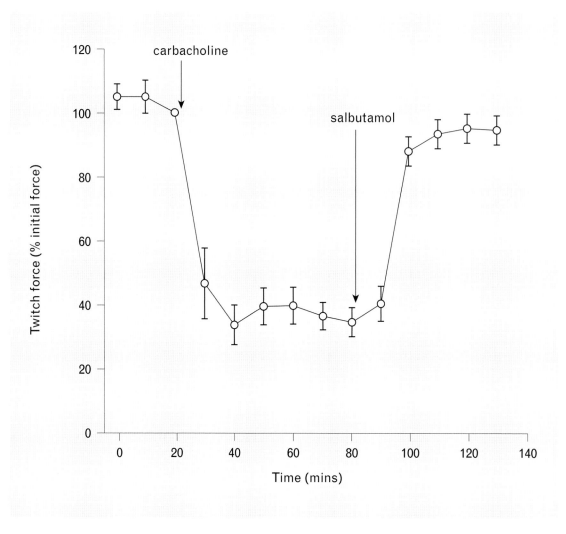

Fig. 26. Time course of the effects of carbacholine (10^{-4} M) and salbutamol (10^{-5} M) on twitch force in iso-lated rat soleus incubated at 30 °C. (Macdonald et al., *J. Physiol.*, 2005).

►► **Fig. 27.** Shows that the stimulating effect of carbacholine on Na^+ influx (A) and Na^+ content (B) is most pronounced in the mid-region close to the motor end plates of the muscle (Macdonald et al., *J. Physiol,* 563:459-469, 2005). Statistical significance is indicated by asterisks and #.

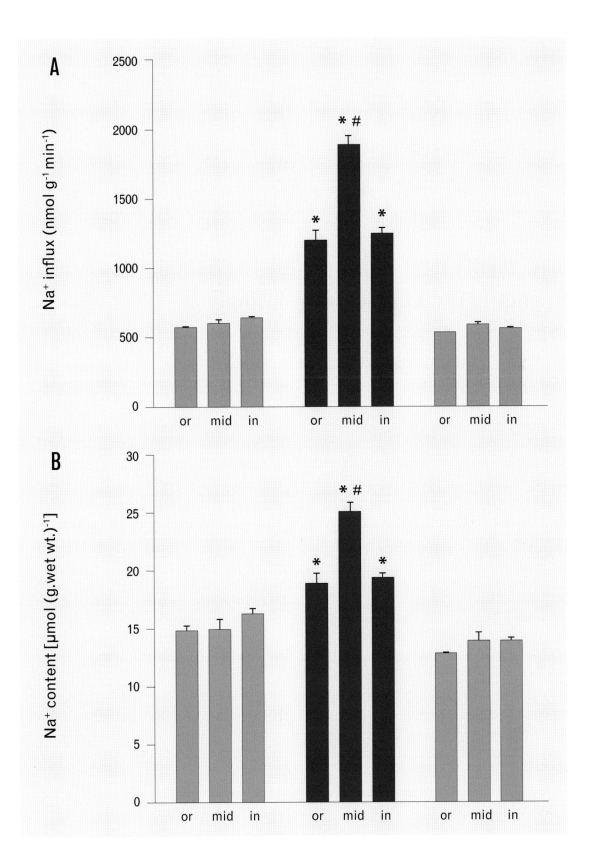

71

17

The importance of Na$^+$-influx in causing depolarization and muscle fatigue

This relationship can be studied 1) by stimulating isolated muscles at increasing frequency. 2) by comparing muscles with predominantly type 1 and type 2 muscle fibres, where excitation-induced Na$^+$ influx is low and high, respectively; 3) by characterizing the changes in Na$^+$ influx and muscle fatigue induced by agents causing increases in Na$^+$-influx. Excitation-induced Na$^+$ influx increases [Na$^+$]$_i$, causes depolarization and fatigue increasing with the frequency of stimulation (Sacco et al, 1994; Clausen et al., 2004).

In rat soleus carbacholine was used for chronic activation of the nicotinic acetyl choline (nACh) receptors to mimic excitation-induced physiological changes in membrane permeability to Na$^+$ and K$^+$ (Macdonald et al., 2005). Carbacholine (10^{-4} M) induced a rapid 2.6-fold increase in Na$^+$ influx, resulting in a 36 % increase in [Na$^+$]$_i$, 13 mV depolarization and reduction in tetanic force to 31 % of controls. This loss of force and the changes in [Na$^+$]$_i$ and [K$^+$]$_i$ and [K$^+$]$_i$ underwent significant recovery upon Na$^+$- K$^+$-pump stimulation with salbutamol, adrenaline, or CGRP. Stimulation of the Na$^+$, K$^+$ pumps allows partial recovery of contractility by restoring excitability through electrogenically driven repolarization of the muscle membrane. As shown in Fig. 26, the considerable suppression of twitch force in rat soleus induced by carbacholine (10^{-4} M) was in 20 min almost completely restored by 10^{-5} M salbutamol. As shown in Fig. 27, the carbacholine-induced Na$^+$-influx and increase in Na$^+$ content in rat soleus was most pronounced in the mid region close to the motor end plates of the muscle (Macdonald et al., 2005).

In conclusion, stimulation of the Na$^+$, K$^+$ pumps allows a muscle to partially recover contractility by regaining excitability through electrogenically driven repolarization of the muscle membrane. It would be interesting to evaluate whether this repolarization might alleviate muscle fatigue.

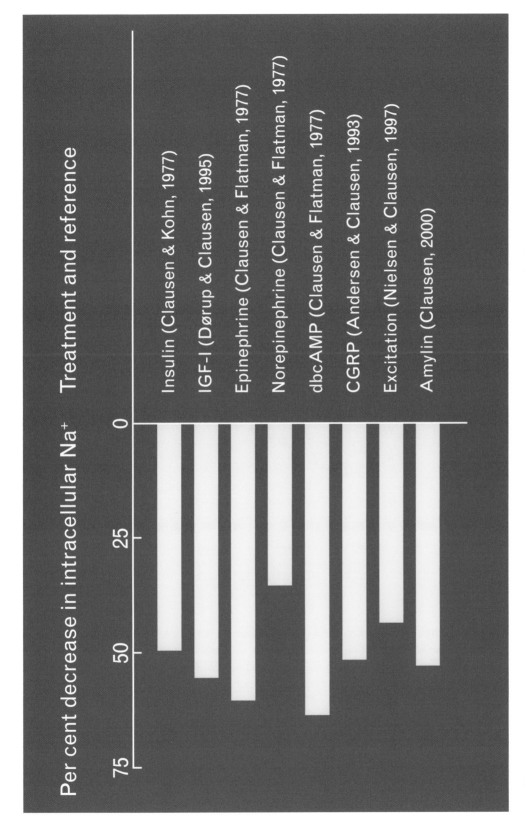

Fig. 28. The stimulating action of various hormones, excitation and dibutyryl cyclic AMP on the Na$^+$, K$^+$-pumps in skeletal muscle all lead to a marked decrease in intracellular Na$^+$. (Clausen Review Lecture, Bratislava, 2007, published in *Acta Physiologica*, vol. 192:351-357, 2007).

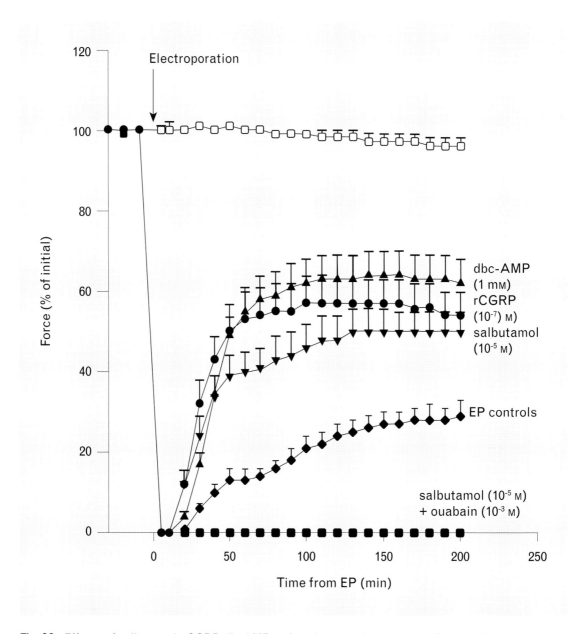

Fig. 29. Effects of salbutamol, rCGRP, dbcAMP and ouabain on the response of tetanic force to electro-poration in rat soleus. After electroporation, force decline and recovery was followed in KR buffer without addditions (n=13) or with ouabain (10^{-3} M, n=4). Each point represents the mean of observations on four to 13 muscles with bars denoting SEM (Clausen & Gissel, *Acta Physiol. Scand.* 2005).

Effect of varying Na⁺, K⁺-pump stimulating agents on intracellular Na⁺ in rat soleus muscles and how this may compensate functional defects caused by plasma membrane leakage

As shown in Fig. 28, stimulation of the Na⁺, K⁺-pumps causes 30-60 % decrease in intracellular Na⁺, documenting the availability of efficient and varying mechanisms for generating and maintaining the transmembrane Na⁺ gradients so essential for muscle cell excitability and symport processes.

The plasma membranes of skeletal muscles may develop non-specific leaks caused by intense work, anoxia, physical damage, electroporation (the formation of pores in the plasma membrane) by exposure to sudden electric currents), excessive swelling due to exposure to hypotonicity, rhabdomyolysis, or various muscle diseases. The response to such leaks and the possible role of the Na⁺, K⁺ pumps in compensating the ensuing functional defects have been examined using electroporation of isolated rat soleus and EDL muscles (Clausen, 2005; Clausen & Gissel, 2005). Muscles were mounted in

an electroporation cuvette and exposed to short (0.1 ms) pulses of an electrical field of 100-800 V/cm across the muscles. This induces rapid formation of pores in the plasma membrane, increasing its permeability, but not that of the membranes of intracellular organelles. This allows reversible influx of Na⁺, loss of K⁺ and excitability, release of intracellular proteins and penetration of extracellular markers into the cytoplasm. As shown in Fig. 29 (Clausen & Gissel, 2005), eight electroporation (0.1 ms) pulses of 500 V/cm induced rapid complete loss of tetanic force in isolated rat soleus muscles, which over the next 200 min was followed by spontaneous force recovery to around 30 % of the force measured before electroporation. The initial rate of this recovery was considerably enhanced (by 183-433 %) and steady-state force level (by 104-143 %) by stimulating the Na⁺, K⁺ pumps with salbutamol,

75

rat CGRP, epinephrine or dibutyryl cAMP, respectively. Both the spontaneous 30 % force recovery and the 50 % force recovery induced by salbutamol were abolished by ouabain, indicating that they were caused by Na^+, K^+-pump stimulation, compensating the leaks of Na^+ and K^+ through the electroporation-induced pores. The electroporation caused depolarization from -70 mV to around -20 mV, followed by a partial spontaneous recovery (Fig. 7 in Clausen & Gissel, 2005). Salbutamol (10^{-5} M) further augmented the repolarization by 15 mV. In isolated rat EDL muscles, 30-60 min of intermittent fatiguing stimulation caused a drop in tetanic force to 12 % of the initial level, followed by slow spontaneous recovery to 20-25 % of the initial force level. (Mikkelsen et al., 2006). This was associated with 11-15 mV depolarization and marked loss of the intracellular protein lactic acid dehydrogenase. Subsequent stimulation of the Na^+- K^+ pumps with salbutamol restored membrane potential to normal level. Salbutamol, epinephrine, rat CGRP, and dibutyryl cAMP all induced a significant increase (40-90 %) in the force recovery after intermittent stimulation (Mikkelsen et al., 2006).

In conclusion, it is interesting that during intensive exercise which may often be associated with bruises and muscle cell damage, the concomitant burst of epinephrine secretion from the adrenals in the intact organism may help restoring the depolarizing effect of such leaks as well as the impairment of excitability. Moreover, such functional impairments might be compensated with injections or inhalations of β_2-agonists.

Long-Term Regulation of Na^+, K^+ Pump Content

As shown in the right side of Fig 16 there are several examples of long-term up- or down- regulation of the content of Na^+, K^+-pumps in skeletal muscle. This has been documented in measurements of [^{3}H]ouabain binding sites in a variety of muscle biopsies from animals and human subjects. Long-term regulation takes place over days to weeks, for instance in working muscles where the synthesis of Na^+, K^+-pumps is significantly augmented (by 14 to 40 % in 8 different studies on the vastus lateralis muscle of human subjects (see table 2 in Clausen, 2003). The training-induced up-regulation of the Na^+,K^+-pump capacity has been detected as reduced duration of exercise-induced hyperkalemia (McKenna et al., 1993) as well as a faster exponential post-exercise decay of plasma K^+ with time-constants of 56 and 98 s in the trained and untrained subjects, respectively (Marcos & Ribas, 1995). This is evidence that training-induced upregulation of Na^+, K^+-pump capacity favours the clearance of extracellular K^+.

Conversely, long-lasting inactivity or immobilization (for instance by paralysis due to nerve damage) leads to downregulation of the contents of Na^+, K^+-pumps that can be measured in muscle biopsies and expressed in pmol/g wet wt. The effects of thyroid hormones, adrenal steroids, fasting, diabetes, K^+-deficiency and K^+ overload will be described below.

In conclusion, quantification of the content of Na^+, K^+-pumps is an important part of the characterization of skeletal muscle performance and endurance.

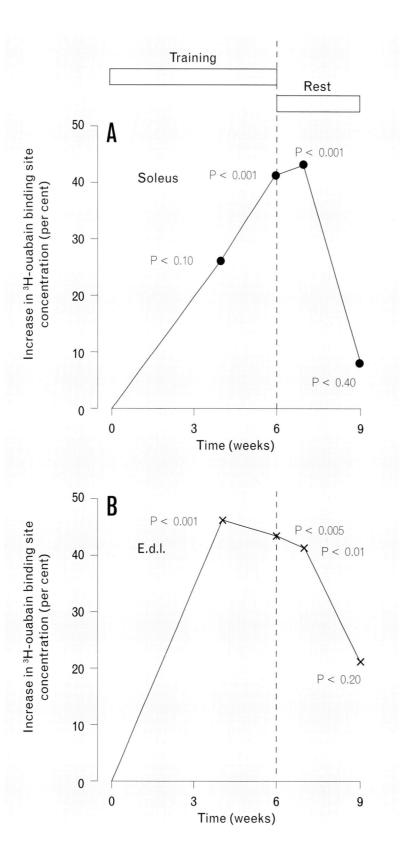

Fig. 30. Time-course of effects of training, rest and detraining on content of [³H]ouabain binding sites in rat soleus and extensor digitorum longus muscles. Rats 6 wks old were trained by daily swimming for 6 wks. and then allowed to rest for up to 3 wks. (Kjeldsen et al, 1986, reprinted by permission of *Biochim. Biophys. Acta*).

Training, inactivity and denervation

Over the last 30 yrs, many longitudinal as well as cross-sectional studies have demonstrated that in a wide variety of species and muscles, training leads to up-regulation and inactivity or denervation to down-regulation of the content of Na^+, K^+ pumps in skeletal muscle

(Table 2 in Clausen 2003, showing the results of 18 different studies, demonstrate that in nine different species, activity induced between 14 and 86 per cent increase in the content of ^3H-ouabain binding sites. Conversely in 5 studies immobilization of skeletal muscle was shown to induce between 20 and 27 per cent decrease in ^3H-ouabain binding sites). The first detection of training-induced upregulation of Na^+, K^+-ATPase enzyme activity was obtained in sarcolemma membranes prepared by differential centrifugation of homogenates of dog gracilis muscle (Knochel et al., 1985). After 6 wk of treadmill training, a maximum increase of 165 % increase was observed. This is far above the upregulation observed in the many later reports on training effects. This discrepancy may be due to the very low and variable recovery (0.2 – 8.9 %) of the total content of Na^+, K^+-ATPase often observed using differential centrifugation (Hansen & Clausen, 1988).

As shown in Fig. 30, swim training for 6 wks at a rather high intensity induced between 22 and 46 % increase (P<0.005 – 0.05) in the total content of [^3H]ouabain binding sites in the intact hind-limb muscles of rats. This was associated with 76 % increase in citrate synthase (P<0.001), but no change in Na^+,K^+ contents, indicating that the active Na^+,K^+-transport could keep pace with the exercise-induced increase in Na^+,K^+-leaks. Since the up-regulation is seen in the muscles participating in the activity and not in other muscles from the same animal, it is not likely to arise from effects of endocrine factors acting on the entire organism (Kjeldsen et al., 1986c; Suwannachot et al., 2000; Suwannachot et al., 2001). Besides, the swim training of the rats caused no significant increase in the thyroid hormone levels (T_3 and T_4) measured in the plasma in the same animals. The up-regulation in [^3H]ouabain binding in the muscles was shown to be reversible, returning to the control level after 3 wks of rest (see Fig. 1 in Kjeldsen et al.,

1986c). Saturation curves showed that the upregulation could not be accounted for by affinity changes in the binding of ^{3}H-ouabain.

In keeping with the training-induced upregulation of Na^+, K^+ pumps in skeletal muscle, hyperventilation induced by chronic hypoxia was found to induce a 24 % increase in the content of Na^+, K^+ pumps in rat diaphragm, which was associated with improved endurance (McMorrow et al., 2011). This is important for patients suffering from hypoxia due to reduced supplies of oxygen or constricted airways. During exposure to cold environment cooling is reduced by muscle shivering, which as detected by electromyography takes place most of the time, the so-called shivering thermogenesis. This continuous muscle activity which is not seen in animals kept at 35 °C leads to rather pronounced upregulation of the content of [^{3}H]ouabain binding sites in the muscles. Thus in the longissimus dorsi muscle of pigs exposure of the animals to 10 °C for 3-4 wks increased the content of [^{3}H]ouabain binding sites by 84 % as compared with pigs kept at 35°C (Harrison et al., 1994).

Denervation reduced the content of [^{3}H]ouabain binding sites in mouse and rat skeletal muscle (Clausen et al., 1981; Clausen et al., 1982; Ward et al., 1987). Immobilization of hind-limbs caused 20-22 % down-regulation of [^{3}H]ouabain binding sites. The entire activity-dependent increase, from immobilized rat muscles to the swim-trained muscles, amounted to 83 %. In guinea pig hindlimbs, immobilization caused a down-regulation of up to 25 % in the content of [^{3}H]ouabain binding sites. In young rats, a similar reduction was measured (Ward

et al., 1987). When subsequently trained by running, the content of [^{3}H]ouabain binding sites in guinea pig hind-limb muscles increased by 57 %. The highest content of [^{3}H]ouabain binding sites in trained control muscles was 93 % above the lowest level measured after 2 wk of immobilization. In another study with sheep, 9 wk of immobilization of a hind-limb caused 22 % decrease in the content of [^{3}H]ouabain binding sites in the vastus lateralis muscle. This loss was restored during a subsequent remobilization (Jebens et al., 1995).

In human subjects, immobilization of the deltoid muscle developed in patients with impingement syndrome of the shoulder (causing marked reduction in mobility) was associated with a significant reduction (27 %) in the content of [^{3}H]ouabain binding sites (Leivseth & Reikeras, 1994). Denervation at the sciatic level was in 7 days found to induce significant reduction (13 and 22 %) in the content of ^{3}H-ouabain binding sites in rat soleus and EDL muscle (Clausen et al., 1982).

Complete interruption of spinal cord was shown to lead to considerable reduction in the content of [^{3}H]ouabain binding sites in human leg muscles (Boon et al., 2012). Recently, knee injury was found to reduce the content of [^{3}H]ouabain binding sites in the vastus lateralis muscle by 20 % in comparison to that of the contralateral muscle (Perry et al., 2015).

In conclusion, the content of Na^+, K^+-pumps in skeletal muscle seems determined by the mobility of the muscles and requires the continued active Na^+, K^+- transport activity of the muscle allowing maintenance of excitability.

Muscular dystrophy and McArdle disease

In homogenates of skeletal muscle samples from patients with myotonic muscular dystrophy, the content of [^{3}H]ouabain binding sites was three to six-fold lower than in control subjects (Desnuelle et al. 1982). The reduced Na$^+$,K$^+$ pump content may explain the observation that in patients with muscular dystrophy, the muscles show increased [Na$^+$]$_i$ and depolarization (Gruener et al., 1979; Edström and Wroblewski, 1989). The reduced Na$^+$, K$^+$ pump content may in part be due to their limited physical activity. However, muscle cells cultured from patients with muscular dystrophy were found to contain 30-40 % fewer [^{3}H]ouabain binding sites than those from age-matched controls (Benders et al., 1996). Obviously, this downregulation cannot be attributed to reduced physical activity. A reduced Na$^+$, K$^+$ pump capacity may contribute to the abnormally high exercise hyperkalemia observed in patients with muscular dystrophy (2.2 vs. the 0.8 mM increase in plasma K$^+$ of controls), (Wevers et al., 1990).

Microelectrode measurements of intracellular Na$^+$ activity in EDL, soleus and gastrocnemius muscles of dystrophic mice showed a marked increase, and intracellular K$^+$ activity was decreased (Ward & Wareham, 1984; Fong et al. 1986). However, the rate of Na$^+$ clearance measured following an excitation-induced Na$^+$ clearance measured following excitation-induced Na$^+$ loading showed no difference between dystrophic and normal mice, indicating that the ability extrude Na$^+$ was not affected (Fong et al., 1986). In conclusion, muscular dystrophy is associated with downregulation of the content of Na$^+$, K$^+$ pumps in skeletal muscle. The ensuing impairment of Na$^+$, K$^+$ homeostasis may contribute to the physical disability of these patients. Patients with McArdle disease suffer from muscle phosphorylase deficiency, which severely restricts the energy supply from glycogenolysis and exercise-induced hyperkalemia is more pronounced (Haller et al., 1998; Paterson et al., 1990). During stimulation of the ulnar nerve at 20 Hz, the amplitude of the compound action potential in the muscles undergoes 50 % decline in 50 s, which is markedly faster than in control subjects. This suggests that membrane excitability is reduced by the elevated [K$^+$]$_o$ (Brandt et al., 1977; Dyken et al., 1967; Griggs et al., 1995). Measurements of [³H]ouabain binding to muscle biopsies showed a significant (27 %) reduction (from 317 to 231 pmol/g wet wt, P<0.0004), which could contribute to the doubling of exercise-induced hyperkalemia observed in the same patients (Fig. 3 in Haller et al., 1998). It has not yet been settled, however, whether the downregulation of Na^+, K^+ pumps reflects a primary reduction in the synthesis of Na^+, K^+ pumps or the reduced physical activity of these patients.

An alternative explanation of the exercise-induced hyperkalemia and accelerated loss of excitability might be inadequate ATP supply from glycolysis to the Na^+, K^+ pump. EDL muscles exposed to anoxia show a considerable loss of force, which can be restored by the β_2 agonists salbutamol or terbutaline (Fredsted et al., 2012). The effect of salbutamol was prevented by blocking the Na^+, K^+ pumps with ouabain or by blocking glycolysis with 2-deoxyglucose. It has been proposed that the Na^+, K^+-pumps in cardiac Purkinje cells (Glitsch & Tappe, 1993) and in t-tubules of skeletal muscle (Dutka & Lamb, 2007) preferentially use ATP from glycolysis.

In conclusion, an advantage of this is that during lack of oxygen, an extra reserve of glycolytic ATP is available to keep the Na^+, K^+-pumps going in one of its major pools, the skeletal muscles, which is most decisive for the maintenance of mobility.

▶▶ **Fig. 31.** Obtained from http://physrev.physiology.org/content/79/4/1317. (Lehmann-Horn and Jurkat-Rott, 1999)
Photo: A paralytic attack in hyperkalemic periodic paralysis elicited by rest after exercise. A naturally occurring animal model, affected quarter horse, is shown. (Courtesy of Dr. E. P. Hoffman.) *Schematic Diagram*: Sequence of cellular events in hyperkalemic periodic paralysis patients that may result in muscle paralysis. $[K^+]_e$, extracellular K^+.

EXPLANATION FOR PARALYTIC ATTACKS IN
HYPERKALEMIC PERIODIC PARALYSIS PATIENTS

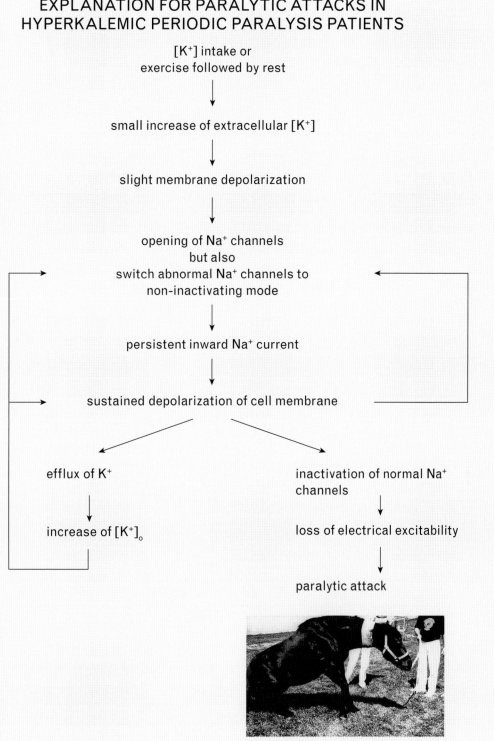

$[K^+]$ intake or
exercise followed by rest

↓

small increase of extracellular $[K^+]$

↓

slight membrane depolarization

↓

opening of Na^+ channels
but also
switch abnormal Na^+ channels to
non-inactivating mode

↓

persistent inward Na^+ current

↓

sustained depolarization of cell membrane

efflux of K^+

↓

increase of $[K^+]_o$

inactivation of normal Na^+
channels

↓

loss of electrical excitability

↓

paralytic attack

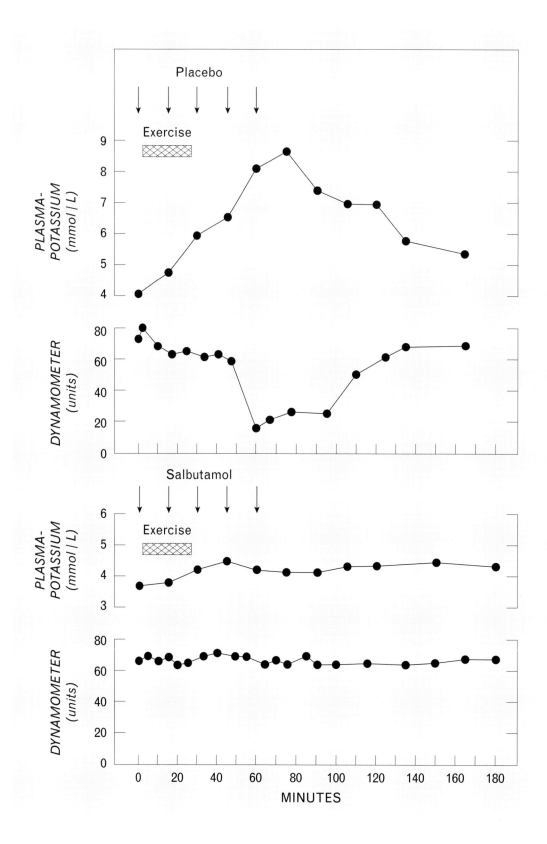

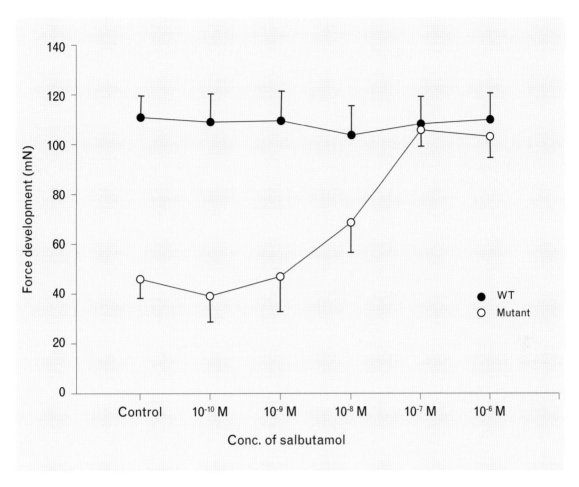

Fig. 33. Effects of salbutamol on tetanic force in soleus obtained from wild typer (WT) and mutant mice. Muscles were stimulated using 2-s 120 Hz trains of 0.2 ms pulses at 7 V in the presence of 10^{-5} M tubocurarine and salbutamol at the concentrations indicated. Each point represents the mean of observations on 4 muscles, with error bars denoting SEM. In the absence of salbutamol, the difference between tetanic force in WT and mutants was highly significant (P<0.001). In the mutant muscles, salbutamol induced a significant increase in tetanic force at 10^{-8} M (P<0.05, 10^{-7} M (P<0.001, and 10^{-6} M (P<0.001). (Clausen et al, *J. Gen. Physiol.* 2011)

◄ ◄ **Fig. 32.** Effect of salbutamol inhalations on plasma–potassium and muscular strength in an attack of hyperkalæmic paralysis provoked by exercise. On 2 consecutive days, the patient exercised on a bicycle ergometer (500 kp m min^{-1} for 30 min) and was then allowed to rest lying down. Each arrow indicates administration of 2 inhalations (i.e., 200 µg) of salbutamol. (Wang & Clausen, *The Lancet*, 1976, p. 211).

Na⁺, K⁺-pump stimulation improves contractility in isolated muscles of mice with hyperkalemic periodic paralysis

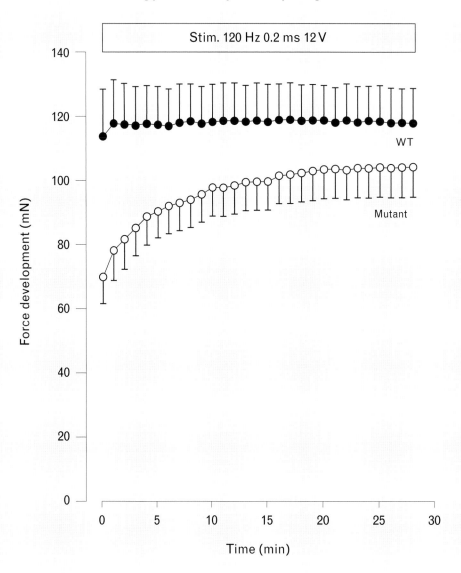

Fig. 34. Effects of repeated stimulation on tetanic force in soleus muscles obtained from 12.5-wk-old wild typer (WT) and mutant mice. Muscles were mounted in force transducers for isometric contractions and stimulated at 120 Hz using 2-s trains of 0.2-ms 12-V pulses every min. Each curve represents the mean of observations every min on four wild type and eight mutant muscles, with error bars denoting SEM. During the initial contractions, the difference between tetanic force in wild type and mutants was significant (P < 0.02). In the mutants, the last tetanic contractions were significant larger than the first (P<0.001). (Clausen et al, *J. of Gen. Physiol.* 2011).

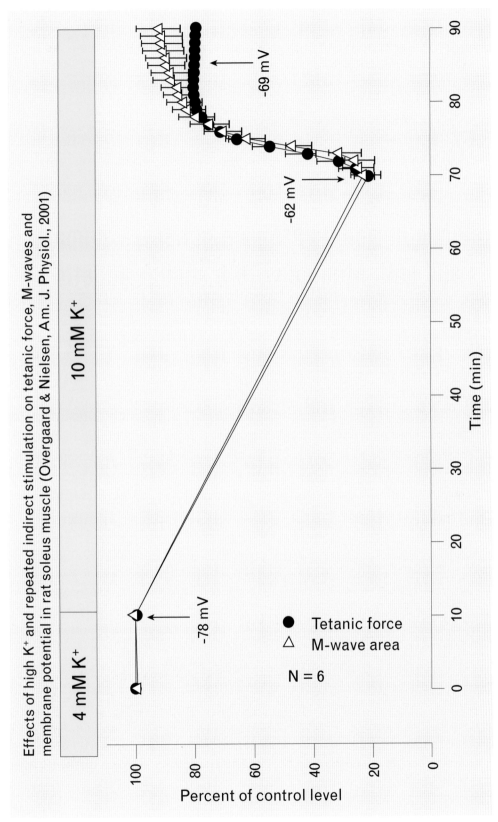

Fig. 35. Excitation-induced recovery of tetanic force and M wave area. Tetanic force and M waves (recorded from the extrajunctional region) measured in soleus muscle in standard KR buffer (4 mM K⁺) and after incubation in buffer with 10 mM K⁺. Tetanic force and M wave area are expressed as percentages of the average value in standard KR buffer. After 60 min (10-70 min) of exposure to 10 mM K⁺ during which the muscles were not stimulated, tetanic stimulations (1.5 s, 30 Hz) were applied at 1 min intervals for 20 min. Symbols represent mean values ± SEM from 6 muscles. Arrows indicate time points where recordings of membrane potential, twitch force and M waves, respectively, in the junctional region were obtained. Overgaard and Nielsen, *Am. J. Physiol.* 2001

Hyperkalemic periodic paralysis in horses, human subjects and mutant mice with a similar genetic anomaly

Fig. 31 shows an American quarterhorse developing an attack of hyperkalemic periodic paralysis as well as a brief explanation of the mechanism of this rather dramatic event. Fig. 32 shows the effects of exercise, followed by rest in a human patient suffering from hyperkalemic periodic paralysis. The time-course of changes in plasma K^+ and muscle force recorded with a hand-dynamometer are shown for a patient treated by 2 inhalations of salbutamol (corresponding to 200 µg) or placebo. Note that when the patient is given placebo, plasma K^+ reaches an alarming level around 8 mM accompanied by complete paralysis of 1 hr duration. In contrast, when the patient is treated with salbutamol, there is almost no change in plasma K^+ or muscle force. This means that these most embarrassing paralytic attacks can be suppressed quickly by inhalation of salbutamol. Later studies have shown that this beta$_2$-agonist can also be given orally or as a subcutaneous injection for the treatment

of hyperkalemic periodic paralysis (Sowinski et al., 2005).

In patients with hyperkalemic periodic paralysis (HPP), a rare inheritable disorder, hitting about one in 100.000 human subjects, attacks of muscle weakness or paralysis are triggered by K^+ ingestion or rest after exercise. Force can within a few minutes be restored by muscle work or treatment with β_2 agonists, given as inhalation or injection (Wang & Clausen, 1976; Clausen et al., 1980). A missense substitution corresponding to a mutation in the skeletal muscle voltage-gated Na^+ channel causing human hyperkalemic periodic paralysis was targeted into the mouse SCN4A gene (mutants). In soleus from these mutants, twitch, tetanic force and endurance were markedly reduced compared with soleus from wild type, reflecting impaired excitability (Hayward et al., 2008). The muscles prepared from these mutants allow a close analysis of the mechanisms of paralysis and its prevention or treatment

(Clausen et al., 2004; Ammar, Lin, Higgins, Hayward and Renaud, 2015). In mutant mouse soleus, contractility is considerably more sensitive to elevated $[K^+]_o$ than wild type soleus.

As shown in Fig. 33 in mutant mice, salbutamol induced significant increase in tetanic force at 10^{-8} M (P<0.05), 10^{-7} M (P<0.001), and (10^{-6} M (P<0.001). In soleus isolated from mutant mice, force development was restored by stimulating the Na^+, K^+-pumps by increasing $[Na^+]_i$ with monensin as well as by the addition of rat CGRP or capsaicin (the active substance in pepper, acting by provoking the release of endogenous CGRP). As shown in Fig. 34, repeated excitation gradually restored contractility, possibly because of the release of endogenous CGRP from stores in the nerve endings in the muscles. These studies on an animal model for hyperkalemic periodic paralysis may explain how mild exercise by triggering the release of CGRP from nerve endings helps locally to prevent severe weakness during an attack of hyperkalemic periodic paralysis (Clausen et al. 2011). It is a returning experience among patients with HPP that during paralytic attacks force can be restored by exercise. In contrast, if mobilization is prevented, paralysis continues, which is most embarrassing, for instance during a bus trip, where a patient may not be able to move up to the bus driver to ask for a required stop of the bus.

The possible mechanism of action of repeated excitation on contraction was also examined in isolated normal rat soleus muscles exposed to high extracellular K^+. As shown in Fig. 35, when the soleus was exposed to KR buffer containing 10 mM K^+, force and M-wave area decreased by about 75 % in 60 min. This was accompanied by a depolarization from 78 mV to 62 mV. When stimulated every min, membrane potential, force and M-wave area were essentially restored in about 10 min (Overgaard and Nielsen, 2001).

In conclusion, these changes are likely to reflect release of CGRP from the nerve endings in the muscles, causing localized stimulation of the Na^+,K^+-pumps and ensuing electrogenic repolarization of the muscle cells (see Fig. 24).

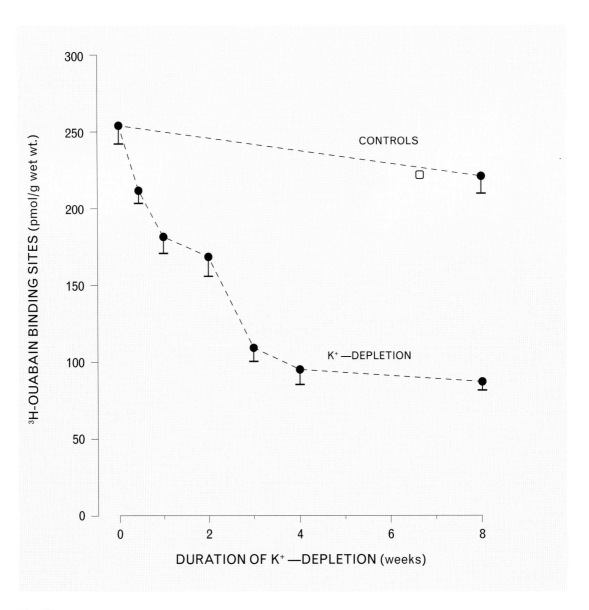

Fig. 36a

Fig. 36a and 36b. The total content of Na^+, K^+-pumps was measured in intact soleus and EDL muscles from young Wistar rats. When maintained on potassium-free fodder, the K^+ content underwent a pronounced decrease in the K^+-content in skeletal muscle, which was associated with a reversible drop in the content of ^3H-ouabain binding sites and a reduced $Na^{+,}$ K^+-pump mediated ^{42}K-uptake and ^{22}Na-efflux. The content of Na^+, K^+-pumps was normalized after 6 days on K^+ repletion. (Nørgaard, Kjeldsen and Clausen, *Nature,* 1981).

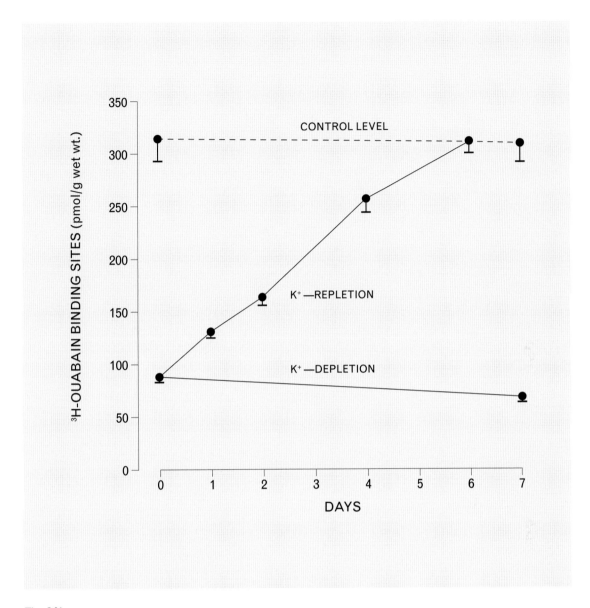

Fig. 36b

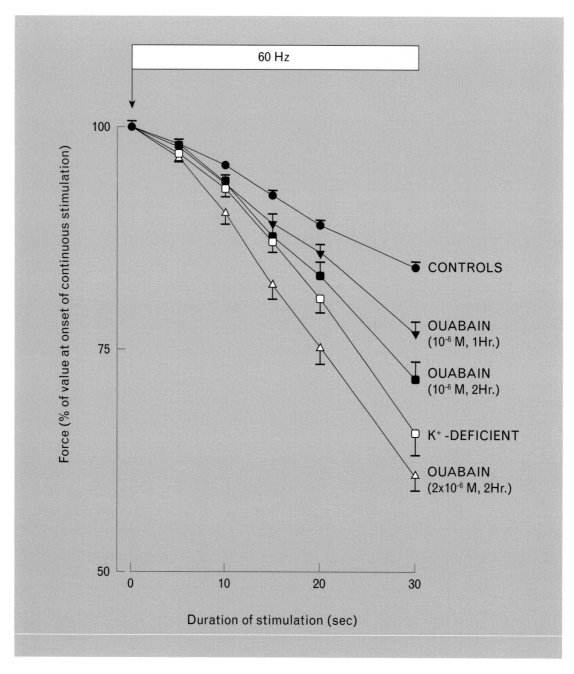

Fig 37. Effects of ouabain or K⁺ deficiency on contractile endurance in rat soleus. Isolated muscles were mounted at optimal length for contraction and exposed to continuous stimulation for 30 s at 60 Hz (1 ms pulses of 10 V). Some muscles were preexposed for the indicated intervals of time to ouabain so as to achieve partial blockade of the Na⁺, K⁺-pumps. Other muscles were prepared from K⁺ deficient rats and therefore contained only 46% of the Na⁺, K⁺-pumps measured in the controls. Each point represents the mean of observations on 8-20 muscles with bars denoting SEM. (Nielsen & Clausen, 1996).

K$^+$ deficiency and K$^+$ restoration

In isolated cells cultured in K$^+$-free buffer or in the presence of ouabain intracellular K$^+$ is decreased and intracellular Na$^+$ increased. This is accompanied by a considerable increase (several fold) in the synthesis of Na$^+$K$^+$ pumps (Pollack et al, 1981). Similar upregulation is seen in cultured muscle cells and may also be obtained by incubation with the plant alkaloid veratridine to increase [Na$^+$]$_i$ (Brodie & Sampson, 1989a,b,c; Wolitzky & Fambrough, 1986). It was surprising, therefore, that in rats maintained on K$^+$ deficient fodder, the content of Na$^+$, K$^+$ pumps of skeletal muscle underwent a marked decrease of up to 80 % (Nørgaard et al., 1981). This downregulation was confirmed by a reduction ^{42}K- uptake in muscles prepared from K$^+$-deficient animals (for details, see Clausen, 1986). Furthermore, isolated rat soleus muscles prepared from K$^+$-deficient rats showed a close correspondence between the reduction of [^{3}H]ouabain binding sites and the reduction in maximum capacity for Na$^+$, K$^+$ pumping (1977). In vivo measurements of [^{3}H]ouabain binding capacity in skeletal muscle showed that K$^+$ deficiency caused the same downregulation as observed when [^{3}H]ouabain binding took place in vitro (Clausen et al. 1982).

K$^+$-deficient rats also showed impaired ability to clear injected [^{3}H]ouabain from the blood plasma. Thus, [^{3}H] activity measured in plasma 15 min after an intraperitoneal injection of [^{3}H]ouabain was 77 % higher in K$^+$-deficient rats than in controls receiving the same dose of [^{3}H]ouabain per kilogram body wt. Clausen et al., 1982). In rat hindlimb muscles 2 wk on K$^+$-deficent fodder decreased the alpha$_2$ isoform by 82 % (Azuma et al., 1991). K$^+$-deficient rats showed a reduced capacity to clear K$^+$ from plasma following oral administration of KCl (Sadre et al, 1987). This has clinical importance because treatment of K$^+$-deficiency with intravenous or oral K$^+$ supplies might cause an extra risk of inducing life-threatening levels of hyperkalemia. Recently, however, a more detailed study showed that in K$^+$-deficient rats, the clearance of intravenously infused K$^+$ was more efficient than in the controls (Bundgaard & Kjeldsen, 2001).

K$^+$-deficiency is one of the most common electrolyte disorders, caused by inadequate dietary supplies of K$^+$. Since Mg^{2+}-deficiency leads to renal loss of K^+, this indirectly also gives rise to K^+ deficiency. Conversely, oral magnesium supplementation restores the concentrations of magnesium, potassium and Na^+,K^+-pumps in skeletal muscle of patients receiving diuretic treatment (Dørup, Skajaa and Thybo, 1993). Other common causes of K^+-deficiency is diarrhea, repeated vomiting, renal insufficiency, treatment with laxatives, diuretics, some antibiotics or the commonly used anticancer drug cisplatinum. Several studies have shown that K^+-deficiency causes loss of Na^+, K^+ pumps from the skeletal muscles, both in animals and in man (Nørgaard et al., 1981; Kjeldsen et al., 1986b; Dørup et al., 1988a; Murphy et al. 2008). In particular in the Third World, K^+-deficiency is very common. Besides, all over the world, millions of patients use diuretics every day for the treatment of hypertension, cardiac or lung diseases (Dørup, Skajaa, Clausen and Kjeldsen, 1988b; Dørup, 1994, doctoral thesis). Alone in Denmark, 350.000 patients have developed this type of K^+-deficiency, leading to muscle weakness and fatigue. As shown in Fig. 36a, Nørgaard et al., 1981), rats exposed to K^+ deficieny by being maintained on K^+-free fodder undergo a rapid initial (62 % in 4 wks) decrease in the content of ^{3}H-ouabain binding sites in muscles, followed by a steady state without much further loss. In contrast, K^+-deficient rats given K^+ supplement undergo a rapid and more than 3-fold increase in ^{3}H-ouabain binding sites, reaching the control level in 6 days; see Fig. 36b. (Nørgaard et al, 1981).

Fig. 37 (Nielsen & Clausen, 1996) shows the time-course of force decline in rat muscles (mean values of between eight and twenty different muscles) exposed to continuous 60 Hz stimulation for 30 s. Obviously, in muscles from K^+-deficient rats, the force decline is about twice as rapid as in the controls which offers an explanation of the fatigue experienced by K^+-deficient patients. A somewhat similar increase in force decline could be obtained by pretreating the muscles from nomal rats with ouabain (10^{-6} M for 1-2 hrs or 2×10^{-6} M for 2 hrs).

It is a well-documented clinical experience, that treatment with diuretics must be combined with supplements of K^+ and Mg^{2+}. In a group of 76 patients who had received diuretics for between 1 and 17 yrs, oral administration of Mg^{2+} for 26 wk restored the contents of K^+, Mg^{2+} and [^{3}H]ouabain binding sites in vastus lateralis muscle towards normal levels (Dørup et al. 1993). Conversely, dietary Mg-deficiency leads to loss of Mg, K^+ and ^{3}H-ouabain binding sites from skeletal muscles, both in animals and man (Dørup, 1994).

In conclusion, during treatment of patients with diuretics, the frequently used combination with potassium is not sufficient, but has to be supplemented with magnesium, which has now gained widespread use.

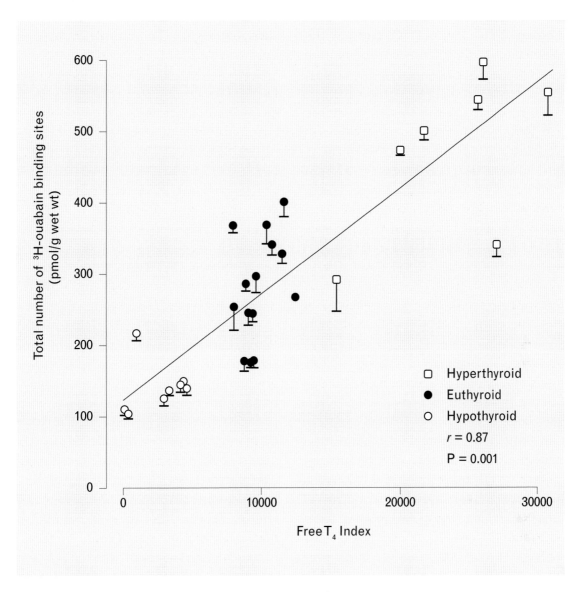

Fig. 38. Relation between free T_4-index and the number of ^{3}H-ouabain binding sites in biopsy specimens of the vastus lateralis muscle of 8 patients with hypothyroidism, 14 euthyroid subjects, and 7 patients with hyperthyroidism. (Kjeldsen et al., *The Lancet*, 1984).

Thyroid hormones and starvation

It is well-established that thyroid hormones increase energy turnover in several tissues. Thus, as shown in (Fig. 1 in Biron et al., 1979), pretreatment of adult mice with thyroid hormone increased basal energy turnover measured as heat production in a microcalorimeter by 46 % (Biron et al., 1979). Conversely, radiation-induced removal of the thyroid gland by injection of [131]I (radiothyroidectomy) or suppression of its secretion by preexposure to perchlorate were shown to induce 32 % reduction in heat production. In all 4 groups of muscles, the Na^+, K^+-pumps were blocked by the addition of a supramaximal concentration of ouabain (10^{-3} M). This was shown to induce a rapid decrease in heat production of 6-8 % (cross-hatched areas of the columns), reaching a new steady state maintained for about 30 min. As shown in (Fig. 2 in Biron et al., 1979), ^{42}K uptake was measured without or with 10^{-3} M ouabain (open or black columns, respectively). As shown in (Table 2 in Biron et al., 1979), the binding of ^{3}H-ouabain was determined as described elsewhere (Clausen & Hansen, 1974). The number of ^{3}H-ouabain binding sites was modified in the same proportion as the ouabain-suppressible ^{42}K uptake. This is in keeping with the original observation that Na^+, K^+-ATPase contents of rat skeletal muscle is increased by thyroid hormone (Asano et al., 1976). In soleus muscle from hyperthyroid rats, the content of [^{3}H]ouabain binding sites was found to be around 10-fold larger than in soleus from hypothyroid rats (Fig. 1 in Kjeldsen, Everts and Clausen, 1986). Similar effects of thyroid status on the content of Na^+, K^+-pumps was also found in human vastus lateralis muscle. Thus, as shown in Fig. 38, the content of [^{3}H]ouabain binding sites increases with thyroid status from 100 to almost 600 pmol/g wet wt., and this increase was closely correlated to the free T_4 index. After therapeutic correction of the hypothyroidism or hyperthyroidism, the content of [^{3}H]ouabain binding sites returned to the level of euthyroid subjects (Kjeldsen et al., 1984a); in a later study, the content of ^{3}H-ouabain binding sites in muscle biopsies from hyperthyroid patients was 89 % (P<0.0001) higher than in those from patients where euthyroidism had been restored by standard methimazol treatment (Riis, Jørgensen, Møller, Weeke and Clausen, 2004).

Studies on cultured rat skeletal myotubes indicate that the thyroid hormone-induced increase in the content of [^{3}H]ouabain binding sites reflects a stimulation of the synthesis of Na$^+$, K$^+$ pumps (Brodie & Sampson, 1988a+b). Cells grown in the presence of T$_3$ or T$_4$ for 24 h showed an increase in [^{3}H]ouabain binding, ouabain-suppressible ^{86}Rb uptake and resting membrane potential. These effects were abolished by blocking protein synthesis. When Na$^+$ influx via Na$^+$ channels or Na$^+$/H$^+$ antiporters were blocked using TTX or amiloride, the thyroid hormone-induced increase in Na$^+$, K$^+$-pump synthesis was suppressed, indicating that it might be elicited by an early increase in Na$^+$ influx. This may be explained by the observation that thyroid hormones increased the content of Na$^+$ channels, as quantifed by [^{3}H] saxitoxin binding by 95 % within 24 h (Brodie & Sampson, 1989a).

In the rat, pretreatment with thyroid hormone induced early increases in resting ^{42}K-efflux, ^{86}Rb efflux and ^{22}Na influx in isolated soleus muscle. These effects were highly significant 12 h after the first injection with T$_3$, clearly preceding the increase in [^{3}H]ouabain binding (Fig. 2 in Everts & Clausen, 1988). Also in rat EDL, a significant stimulation of ^{86}Rb efflux was seen 12h after injection of T$_3$. Conversely, in muscles from hypothyroid rats, ^{86}Rb uptake was decreased by 30 %.

The increase in the content of Na$^+$, K$^+$ pumps was found to be preceded by an upregulation of the content of Na$^+$ channels (Harrison & Clausen, 1998). In rat soleus, the content of [^{3}H]ouabain binding sites and ^{42}K uptake were decreased by 61 % in hypothyroidism and increased by 80 % in hyperthyroidism (Matsumura et al., 1992). Total fasting decreased [^{3}H]ouabain binding to the same level as in hypothyroid rats

(Kjeldsen et al., 1986b). These changes were associated with similar relative changes in ouabain-suppressible ^{42}K uptake. Refeeding or T$_3$ treatment of the fasting rats restored [^{3}H]ouabain binding in skeletal muscle to the normal level.

In rat soleus, hypophysectomy decreased the content of [^{3}H]ouabain binding sites in rat soleus by around 80 %. Subcutaneous injections of T$_4$ at doses restoring the serum concentrations of T$_3$ and T$_4$ to the physiological levels increased the content of [^{3}H]ouabain binding sites in the muscles of the hypophysectomized rats to control level in 11 days (Everts et al. 1990). Human growth hormone at doses causing normalization of serum IGF-I induced no increase in the content of [^{3}H]ouabain binding sites. These observations indicate that the effect of thyroid hormones on the synthesis of Na$^+$, K$^+$ pumps in skeletal muscle is excerted at physiological concentrations and does not depend on growth hormone or IGF-I.

In conclusion, thyroid hormones in the physiological range are the major endocrine factors controlling the content of Na$^+$, K$^+$-pumps in skeletal muscle. Their marked stimulating effect on the synthesis of Na$^+$, K$^+$ pumps seems to be driven by early increases in the passive leaks to Na$^+$ and K$^+$ via Na$^+$ channels and K$^+$ channels, respectively (Everts & Clausen, 1988). This is important for whole body Na$^+$, K$^+$ homeostasis and contractile performance.

Starvation leads to a decrease in the plasma levels of thyroid hormones in humans (Vagenakis et al., 1975) and in rats (Kjeldsen et al., 1986a; Schussler & Orlando, 1978). Complete fasting for 5 days induced a 50 % decrease in the content of Na$^+$, K$^+$-ATPase of sarcolemma prepared from rat skeletal muscle Swann, 1984). Measurements on intact rat muscle samples showed that 5 days of starvation only caused a 25 % decrease in the content of [³H]ouabain binding sites (Kjeldsen et al., 1986b). Reducing the caloric intake to half the normal for 3 wks caused 45 and 53 % decrease in plasma T_3 and T_4, respectively, and a 25 % drop in the content of [³H]ouabain binding sites in soleus muscle. A later study confirmed the downregulation induced by semistarvation (Schmidt et al. 1994), which was also seen in various other types of skeletal muscle. The downregulation of Na^+, K^+ pumps was fully reversible after 3 days of refeeding (Kjeldsen et al., 1986b). The downregulation could not be attributed to K^+ deficiency and may be accounted for by the reduction in thyroid status. Because reduced caloric intake also decreases the total mass of skeletal muscles, the total muscle pool of Na^+,K^+ pumps is severely decreased (in the rat by up to 58 %), leading to impairment of K^+ tolerance. In rats, fasting for 2 days induced a 33 % decrease in the Na^+, K^+ ATPase activity in skeletal muscle (Nishida et al., 1992). Four days of starvation decreased plasma T_3 and the content of [³H]ouabain binding sites in the soleus of 5-wk-old rats to the same value as in hypothyroid rats, effects that were completely restored to the control level by 5 days of refeeding or prevented by concomitant administration of T_3 (Matsumura et al. 1992). Ouabain-suppressible ^{42}K uptake showed similar relative changes as

[³H]ouabain binding. A more recent study on Shetland ponies showed that food restriction causing 30-50 % reduction in body weight gain produced significant decreases in serum T_3 and free T_3 (30 and 49 %, respectively), but only a minor (12-15 %) nonsignificant decrease in the content of [³H]ouabain binding sites in the gluteus medius muscle (Suwannachot et al., 2000). The downregulation of Na^+, K^+ pumps seen during reduced caloric intake might in part reflect reduced physical activity, although it is more rapid than that seen during inactivity. However, it is unlikely to be due to reductions in fiber size, which would rather cause an increase leading to underestimation of the downregulation (Harrison et al, 1994).

In conclusion, globally the downregulation of Na^+, K^+ pumps in skeletal muscle elicited by reduced caloric intake may represent the most commonly occurring Na^+, K^+ pump disorder, important for the ability to tolerate sudden major K^+ ingestion as well as for muscle endurance. Reduced caloric intake may often be combined with K^+-deficiency, caused by reduced K^+ intake, diarhoea, vomiting or treatment with diuretics. Still, there are few specific studies on these anomalies in human subjects. Moreover, it seems likely that reduced food intake in relation to cancer, infections, anorexia, and other diseases might also lead to downregulation of Na^+, K^+ pumps in skeletal muscle.

Thermogenic actions of thyroid hormones and malignant hyperthermia (MH)

The idea that active cation transport is a pacemaker for energy metabolism is often supported by the fact that some tissues contain high concentrations of Na^+,K^+-ATPase (brain, kidney) or Ca^{2+} ATPase (skeletal muscle, heart) and therefore have the capacity to split a major part of the ATP produced. The energy utilization of active Ca^{2+} transport in resting skeletal muscle has been estimated not to exceed a few per cent of the total amount. During contractile activity, however, 20-50 % of the energy is utilized for for Ca^{2+} reaccumulation into the SR. During muscular activity this may contribute to 20-40 % of the thermogenic action of thyroid hormones (Van Hardeveld & Clausen, 1984; Simonides & Van Hardeveld, 1988). The significance of Ca^{2+} ions in the control of energy turnover is well-illustrated by malignant hyperthermia (MH), where a rapid and sometimes lethal rise in body temperature (often 1 $^{\circ}$C every 5 min) is elicited by an increase in free cytoplasmic Ca^{2+} in skeletal muscle.

In conclusion, such attacks may be prevented by blocking the release of Ca^{2+} from the SR, using dantrolene. In conclusion, therefore, for safety reasons, this pharmaceutical must be kept ready for use in clinical departments performing general anaesthesia. The incidence of fulminant MH may vary between 1 in 20,000 and 1 in 200,000 persons subjected to anaesthesia (for details, see Clausen et al., 1991).

26

Diabetes

In rats made diabetic by streptozotocin treatment, the content of [^{3}H]ouabain binding sites in skeletal muscle as measured after 4 wk was reduced by between 24 % (in EDL) and 48 % (in soleus). These changes could be completely prevented by insulin treatment (Kjeldsen et al., 1987). Moreover, in the diabetic rats, 8 wk of treatment with insulin increased the content of [^{3}H]ouabain binding sites in skeletal muscle by 23 %. In rats made diabetic by streptozotocin or partial pancreatectomy, the content of [^{3}H]ouabain binding sites in soleus was reduced by 12-21 % (Schmidt et al., 1994). Again, insulin treatment of streptozotocin-diabetic rats increased the content of [^{3}H]ouabain binding sites to 18-26 % above controls. Moreover, in insulin-treated patients with non-insulin-dependent diabetes mellitus, where plasma insulin was increased, the content of [^{3}H]ouabain binding sites in vastus lateralis was increased by 20 % compared with healthy controls and there was a significant correlation between plasma insulin and the content of [^{3}H]ouabain binding sites in vastus lateralis muscle (Schmidt et al., 1994). In skeletal muscle prepared from rats 2

and 14 days after the induction of diabetes with streptozotocin, the activity of Na^+,K^+-ATPase was decreased by 30-50 % (Nishida et al., 1992). This may explain the observation that in diabetic rats, $[Na^+]_i$ in skeletal muscle is increased (Moore et al., 1983). In contrast, crude membrane preparations obtained from skeletal muscles of streptozotocin treated rats showed a clear increase in the contents of both α_1- and α_2-subunit isoforms, which could be partially reversed by insulin treatment (Ng et al., 1993). The same study showed that in the diabetic heart, the content of α_2-isoform was significantly decreased, and this was partially restored by insulin treatment. These discrepancies remain unexplained. In juvenile human diabetic subjects, the withdrawal of insulin treatment led to a significant drop in the nightly peaks of serum thyroid stimulating hormone (TSH) (46 %) and free serum T_3 (28 %) (Schmitz, Alberti et al., 1981), suggesting that insulin increases thyroid hormone levels via an effect on pituitary function. In rats, streptozotocin--induced diabetes was associated with a marked reduction in the plasma thyroid hormone level, which could be completely restored by insulin (Sundarean et al., 1984; Zhang et al. 2002).

In conclusion, these observations indicate that insulin stimulates the synthesis of Na$^+$-K$^+$ pumps in skeletal muscle, possibly indirectly by increasing the secretion of thyroid hormones. This might explain the downregulation of muscular Na$^+$, K$^+$ pump content seen in untreated diabetes.

Steroid hormones, Glucocorticoids, Aldosterone

Adrenal steroids influence the synthesis of Na^+, K^+ ATPase in several tissues, but almost no studies have explored their effects on skeletal muscle. However, in an examination of patients with chronic obstructive lung disease (COLD) in intensive treatment with the glucocorticoid dexamethasone it was unexpectedly observed that in biopsies from the vastus lateralis muscle the content of ^{3}H-ouabain binding sites was significantly (P<0.001) increased by 31-61 % (Ravn & Dørup, 1997). A follow-up on these observations showed that in rats treatment with dexamethasone infused via osmotic minipumps (0.02-0.05 mg x kg body wt^{-1} x day^{-1} for 2 wks) increased the content of ^{3}H-ouabain binding sites in soleus and EDL muscles by 34 % and 52 %, respectively (P<0.001). The content of ^{3}H-ouabain binding sites in rat muscles was not reduced after adrenalectomy. Thus, an unstimulated level of endogenous glucocorticoids is of minor importance for the physiological regulation of Na^+, K^+-pumps in skeletal muscle. The stimulating effect of dexamethasone on the synthesis of Na^+, K^+ pumps could not be attributed to mineralocorticoid actions of the compound. Thus aldosterone induced a highly significant downregulation of the content of [^{3}H]ouabain binding sites in all four rat muscles examined (soleus, EDL, gastrocnemius and diaphragm). This could be explained by the concomitant K^+ deficiency caused by aldosterone (Dørup & Clausen, 1997), in keeping with our earlier study showing that fluorohydrocortisone decreases the content of K^+ and [^{3}H]ouabain binding sites in skeletal muscle (Kjeldsen et al., 1984c). A recent study of inhalation of the synthetic glucocorticoid budesonide showed that in ten 18-40 years old men 2 weeks of daily inhalations (4 times 0.4 mg, equivalent to the maximum recommended dose for the treatment of asthma), the content of ^{3}H-ouabain binding sites in biopsies from the vastus lateralis muscle had increased by 17 % (P<0.01). The concentration of ^{3}H-ouabain in the muscle biopsies showed significant correlation to the plasma concentration of budesonide, showing that the standard dose of the glucocorticoid also stimulate the synthesis of $Na,^+$ K^+ pumps in man in vivo (Hostrup et al., 2016).

In conclusion, high doses of glucocorticoids stimulate the synthesis of Na^+, K^+

pumps in skeletal muscle. In a wide variety of disorders large doses of glucocorticoids or ACTH are frequently used for treatment. Thus it can be expected that such patients undergo an upregulation of the content of Na$^+$, K$^+$-pumps in skeletal muscles, which might be of importance for K$^+$ homeostasis, contractile endurance and doping. In asthmatics, the frequently used combination of treatment with the glucocortocoid dexamethasone and the hypokalemia-inducing β$_2$-agonists may augment the risk of more marked hypokalemia.

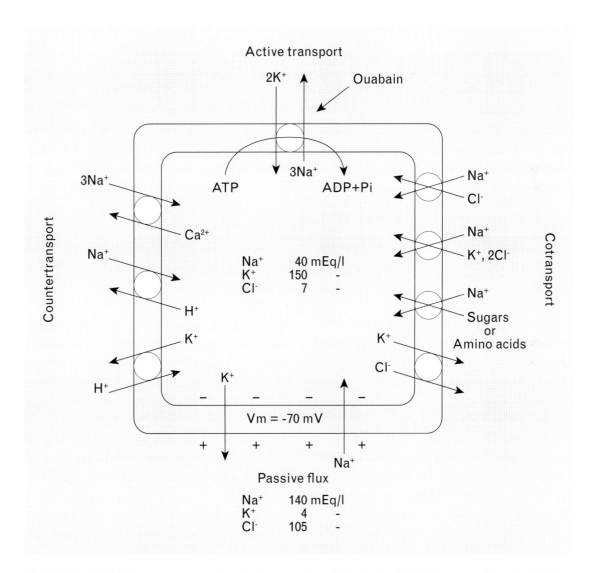

Fig. 39. The Na$^+$, K$^+$ pumps convert chemical energy from hydrolysis of ATP into a gradient for Na$^+$ into and for K$^+$ out of the cell. These gradients are used as energy sources for co- and counter-transport of other substances (Na$^+$,H$^+$, Ca^{2+},Cl$^-$ sugars, amino acids, for polarization of the membrane and for de- or re-polarization of the membrane potential. (Skou, *News in Physiol. Sciences*, 1992).

28

Major Conclusions and General Perspectives

The physiological and clinical significance of the Na^+, K^+ pumps in the skeletal muscles is perhaps best documented by the following findings:

1. The Na^+,K^+ pumps are located in sarcolemma with a large fraction in t-tubular membranes, in optimal close vicinity to Na^+ and K^+ channels.

2. The content of Na^+, K^+ pumps has repeatedly been quantified and their total transport capacity for Na^+ and K^+ found to agree with the theoretical maximum predicted by the content of $[^3H]$ouabain binding sites.

3. Excitation is induced by rapid influx of Na^+, immediately followed by an almost equivalent efflux of K^+. This causes the action potential and within seconds activates the Na^+,K^+ pumps up to 22-fold in rat soleus and 7-fold in rat EDL, sufficient to lead to a subsequent undershoot in $[Na^+]_i$.

4. During high frequency excitation (e.g. 120 Hz) the active Na^+, K^+ transport in intact rat soleus may reach its theoretical maximum rate in 10 s.

5. A wide variety of hormones and pharmaceuticals (insulin, insulin-like growth Factor I, β_2-agonists, calcitonin-gene related peptide, amylin and ATP) induce acute stimulation of the activity of Na^+, K^+ pumps in isolated muscles as well as in vivo. This stimulation may induce repolarization and force recovery, important for rapid restoration of muscle function after intense exercise

6. In isolated intact muscles inhibited by high $[K^+]_o$ low $[Na^+]_o$ or depolarization, Na^+, K^+ pump stimulation induced by excitation, hormones or elevated temperature elicits rapid restoration of excitability.

7. Training upregulates and inactivity or paralysis down-regulates the content of Na^+, K^+ pumps in animal and human muscles. These changes are reversible and associated with increase and decrease in contractile performance, respectively.

8. Inhibition or downregulation of the Na^+, K^+ pumps increases the leak-to-pump ratio and reduces contractility of isolated muscles. Likewise, increasing leak-to

pump ratio by augmenting the activity or content of Na^+ or K^+ channels also reduces endurance, as for example in hyperthyroidism.

9. Several diseases (hypothyroidism, K^+ deficiency, muscular dystrophy, McArdle disease, diabetes and starvation) are associated with downregulation of the content of Na^+, K^+ pumps in skeletal muscle, which is likely to contribute to the loss of physical performance seen in these disorders.

10. The excitation of skeletal muscles causes a rapid and marked drop in $[Na^+]_o$ and increase in $[K^+]_o$. These changes contribute to a substantial part of muscle fatigue, which may therefore be termed Na^+, K^+ fatigue. Excitation seems to be a self-limiting process, restored by rapid stimulation of the Na^+, K^+ pumps.

In sum, the Na^+, K^+-pumps convert chemical energy from ATP hydrolysis into a gradient for Na^+ into and for K^+ out of the muscle cell. As illustrated in Fig. 39, borrowed from a late review on the Na^+, K^+ pump (Skou, 1992), these gradients are used as energy sources for co- and counter-transport of other substances, (Cl^-, Ca^{2+}, H^+, PO_4, sugars, amino acids and pharmaceuticals), for polarization of the membrane and for de- and repolarization of the membrane potential.

References

Albert C M, Mittleman MA, Chae CU, I.-Min Lee, Hennekens CH and Manson, JA E. Triggering of Sudden Death from Cardiac Causes by Vigorous Exertion. *N Engl J Med,* 343:1355-1361, 2000.

Almers W. Potassium concentration changes in the transverse tubules of vertebrate skeletal muscle. *Federation Proc* 39: 1527-1532, 1980.

Ammar, T, Lin**, W, Higgins, A, Hayward, LJ and Renaud, J-M.** Understanding the physiology of the asymptomatic diaphragm of the M1592V hyperkalemic periodic paralysis mouse. *J Gen Physiol* vol. 146 no. 6, 509-525, 2015.

Andersen SLV and Clausen T. Calcitonin gene-related peptide stimulates active Na^+-K^+ transport in rat soleus muscle. *Am J Physiol Cell Physiol* 264: C419-C429, 1993.

Appelt D, Buenviaje B, Champ C, and Franzini-Armstrong F. Quantitation of "junctional feet" content in two types of muscle fiber from hind limb muscles of the rat. *Tissue Cell* 21: 783-794,1989.

Armstrong ML, Dua AK, and Murrant CL. Potassium initiates vasodilatation induced by a single skeletal muscle contraction in hamster cremaster muscle. *J. Physiol.* 581: pp. 841-852, 2007.

Asano Y, Liberman UA, and Edelman IS. Thyroid thermogenesis. Relationships between Na^+-dependent respiration and Na^+ K^+-ade-nosine triphosphatase activity in rat skeletal muscle. *J Clin Invest* 57: 368-379, 1976.

Aughey, RJ, Gore, CJ, Hahn, AG, Garnham, AP, Clark, SA, Petersen, AC, Roberts, A D, McKenna, MJ. Chronic intermittent hypoxia and incremental cycling exercise independently depress muscle in vitro maximal Na^+-K^+-ATPase activity in well-trained athletes. *Journal of Applied Physiology*, Vol. 98:pp. 186-192, 2005.

Azuma KK, Hensley CB, Tang MJ, and McDonough AA. Hypokalemia decreases Na^+-K^+-ATPase α_2- but not α_1-isoform abundance in heart, muscle, and brain. *Am J Physiol Cell Physiol*, 260:C958-C964, 1991.

Ballanyi K and Grafe P. Changes in intracellular ion activites induced by adrenaline in human and rat skeletal muscle. *Pflügers Arch* 411: 283-288, 1988.

Balog EM and Fitts RH. Effects of fatiguing stimulation on intracellular Na^+ and K^+ in frog skeletal muscle. *J Appl Physiol* 81:679-685, 1996.

Bangsbo J, Gunnarson T P, Wendell, J. Nybo L, and Thomassen, M. Reduced volume and increased training intensity elevate muscle Na^+-K^+ pump α_2-subunit expression as well as short- and long-term work capacity in humans. *J Appl Physiol* 107: 1771-1780, 2009.

Benders AAGM, Wevers RA, and Veerkamp JH. Ion transport in human skeletal muscle cells: disturbances in myotonic dystrophy and Brody's disease. *Acta Physiol Scand* 156: 355-367, 1996.

Benowitz, NL, Osterloh J, Goldschlager N, Kaysen, G, Pond, S and Forhan, S. Massive Catecholamine Release From Caffeine Poisoning. *JAMA,* 248(9):1097-1098, 1982

Bigland-Ritchie B, Jones DA, and Woods JJ. Excitation frequency and muscle fatigue: electrical responses during human voluntary and stimulated contractions. *Exp Neurol* 64: 414-427, 1979.

Biron R, Burger A, Chinet A, Clausen T, and Dubois-Ferrie'ere R. Thyroid hormones and the energetics of active sodium-potassium transport in mammalian skeletal muscles. *J Physiol (Lond)* 297: 47-60, 1979.

Bismuth C, Gaultier M, Conso F, Efthymiou ML. Hyperkalemia in acute digitalis poisoning: prognostic significance and therapeutic implications. Clin Toxicol. 6(2):153-62, 1973

Boon H, Kostovski E, Pirkmajer S, Song M, Lubarski I, Iversen PO, Hjeltnes N, Widegren U, Chibalin AV. Influence of chronic and acute spinal cord injury on skeletal muscle Na^+-K^+-ATPase and phospholemman expression in humans. Am J Physiol. 2012, 302:E864-71.

Bouclin R, Charbonneau E, and Renaud RM. Na^+ and K^+ effect on contractility of frog sartorius muscle: implication for the mechanism of fatigue. *Am J Physiol Cell Physiol* 268: C1528-C1536, 1995.

Brandt NJ, Buchtal F, Ebbesen F, Kamieniecka Z, and Krarup C. Post-tetanic mechanical tension and evoked action potentials in McArdle's disease. *J Neurol Neurosurg Psychiatry* 40: 920-925, 1977.

Brodie C and Sampson SR. Effects of carbamylcholine on membrane potential and Na-K pump activity of cultured rat skeletal myotubes. *Cell Mol Neurobiol* 8:393-410, 1988a.

Brodie C and Sampson SR. Characterization of thyroid hormone effects on Na-K pump and membrane potential of cultured rat skeletal myotubes. *Endocrinology* 123: 891-897, 1988b.

Brodie C and Sampson SR. Characterization of thyroid hormone effects on Na^+ channel synthesis in cultured skeletal myotubes: role of Ca^{2+}. *Endocrinology* 125: 842-849, 1989a.

Brodie C and Sampson SR. Regulation of the sodium-potassium pump in cultured rat myotubes by intracellular sodium ions. *J Cell Physiol* 140: 131-137, 1989b.

Brodie C and Sampson SR. Characterization of resting membrane potential and its electrogenic pump component in cultured chick myotubes. *Int J Devl Neurosci* 7: 165-172, 1989c.

Brooks S, Nevill ME, Meleagros L, Lakomy HK, Hall GM, Bloom SR, and Williams C. The hormonal responses to repetitive brief maximal exercise in humans. *Eur J Appl Physiol Occup Physiol* 60: 144-148, 1990.

Buchanan R, Nielsen OB, and Clausen T. Excitation- and β_2- agonist-induced activation of the Na^+, K^+ pump in rat soleus muscle. *J Physiol (Lond)* 545: 229-240, 2002.

Bundgaard H. and Kjeldsen K. Potassium depletion increases potassium clearance capacity in skeletal muscles in vivo during acute repletion. *Am J Physiol Cell Physiol* 283: C1163-C1170, 2001.

Buur, T. Clausen, T. Holmberg, E. Johansson, U. and Waldeck. B. Desensitization by terbutaline of β-adrenoceptors in the guinea-pig soleus muscle: Biochemical alterations associated with functional changes. *Brit. J. Pharmacol.* Vol. 76, pp. 313-317, 1982.

Cairns, SP, Buller, SJ Loiselle DS, Renaud JM, Changes of action potentials and force at lowered $[Na^+]_o$ in mouse skeletal muscle: implications for fatigue. *Am. J. Physiol.* vol. 285:C1131-C1141, 2003.

Cartana J. and Stock, M. Effects of clenbuterol and salbutamol on tissue rubidium uptake in vivo, *Metabolism,* vol. 44: pp. 119-125, 1995.

Castronuovo G, Lippe C, Bellantuono V, Calzaretti G, and Ardizzone A. Effect of calcitonin gene-related peptide on sodium absorption through isolated skin of *Rana esculenta. Arch Physiol Biochim* 104: 142-147, 1996.

Chibalin AV, Kovalenko MV, Ryder JW, Feraille E, Walberg- Henriksson H, and Zierath JR. Insulin- and glucose-induced phosphorylation of the Na^+, K^+-adenosine triphosphatase alpha subunits in rat skeletal muscle. *Endocrinology* 142: 3474-3482, 2001.

Clausen T. The relationship between the transport of glucose and cations across cell membranes in isolated tissues. II. Effects of K^+-free medium, ouabain and insulin upon the fate of glucose in rat diaphragm. *Biochim Biophys Acta* 120: 361-368, 1966.

Clausen T. Regulation of active Na^+-K^+ transport in skeletal muscle. *Physiol Rev* 66: 542-580, 1986.

Clausen T. Long- and short-term regulation of the Na^+-K^+ pump in skeletal muscle. *News Physiol Sci* 11: 24-30, 1996a.

Clausen T. The Na^+,K^+ pump in skeletal muscle: quantification,regulation and functional significance, invited review. *Acta Physiol Scand* 156: 227-235, 1996b.

Clausen T. Clinical and therapeutic significance of the Na^+, K^+ pump, editorial review. *Clin Sci* 95: 3-17, 1998.

Clausen T. Effects of amylin and other peptide hormones on Na^+K^+ transport and contractility in rat skeletal muscle. *J Physiol (Lond)* 527: 121-130, 2000.

Clausen T. The sodium pump keeps us going. Invited review. *Proc NY Acad Sci.* 986: 595-602, 2003a.

Clausen T. Na^+, K^+ pump regulation and skeletal muscle contractility. *Physiol. Rev.* 83: 1269-1324, 2003b.

Clausen T. Na^+-K^+ Pump Stimulation Improves Contractility in Damaged Muscle Fibers. *Annals of the New York Academy of Sciences,* Volume 1066, Cell Injury: Mechanisms, Responses, and Repair, pages 286-294, December 2005.

Clausen T. Role of Na^+,K^+-pumps and transmembrane Na^+, K^+-distribution in muscle function. The FEPS Lecture – Bratislava 2007. Published in *Acta Physiol.* Vol 192: 339-349, 2008a.

Clausen T. Clearance of extracellular K^+ during muscle contraction – roles of membrane transport and diffusion. *J. Gen. Physiol.* vol. 131: 473-481, 2008b.

Clausen T. Hormonal and pharmacological modification of plasma potassium homeostasis. *Fundamental and Clinical Pharmacology*, vol. 24:595-605, 2010.

Clausen T., Excitation-induced exchange of Na^+, K^+ and Cl^- in rat EDL in vitro and in vivo: physiology and pathophysiology. *J. Gen. Physiol.* 141:179-192, 2013a.

Clausen T. Quantification of Na^+, K^+ pumps and their transport rate in skeletal muscle: Functional significance. Invited review. *J. Gen. Physiol.* Vol 142:327-345, 2013b.

Clausen T. Excitation of skeletal muscle is a self-limiting process, due to run-down of Na^+, K^+ gradients, recoverable by stimulation of the Na^+, K^+-pumps. *Physiological Reports*, vol. 3, pp. 1-11, 2015.

Clausen T., Andersen SLV, Flatman JA. Na^+-K^+-Pump Stimulation Elicits Recovery of contractility in paralysed rat muscle. *J Physiol* 472: 521-536, 1993.

Clausen T., Overgaard, K. and Nielsen, O.B. Evidence that the Na^+, K^+ leak/pump ratio contributes to the difference in endurance between fast- and slow-twitch muscles. *Acta Physiol. Scand.* 180: 209-216, 2004.

Clausen T., Overgaard, K., Nielsen, O.B. Clausen, J. D, Pedersen, T.H. and Hayward, L. J. Na^+, K^+-pump stimulation improves contractility in isolated muscles of mice with hyperkalemic periodic paralysis. *J. Gen. Physiol.* 138:117-130, 2011.

Clausen T and Everts ME. K^+-induced inhibition of contractile force in rat skeletal muscle: role of active Na^+-K^+ transport. *Am J Physiol Cell Physiol* 261: C799-C807, 1991.

Clausen T, Everts ME, and Kjeldsen K. Quantification of the maximum capacity for active sodium-potassium transport in rat skeletal muscle. *J Physiol (Lond)* 388: 163-181, 1987.

Clausen T and Flatman JA. The effect of catecholamines on Na^+-K^+-transport and membrane potential in rat soleus muscle. *J Physiol (Lond)* 270: 383-414, 1977.

Clausen T and Flatman JA. β_2-Adrenoceptors mediate the stimulating effect of adrenaline on active electrogenic Na^+-K^+-transport in rat soleus muscle. *Br J Pharmacol* 68: 749-755, 1980.

Clausen T and Flatman JA. Effects of insulin and epinephrine on Na^+ K^+ and glucose transport in soleus muscle. *Am J Physiol* 252: E492-E499, 1987.

Clausen T. and Gissel H. Role of Na^+, K^+ pumps in restoring contractility following loss of cell membrane integrity in rat skeletal muscle *Acta Physiol. Scand.* 183: 263-271, 2005.

Clausen T and Hansen O. Ouabain binding and Na^+-K^+ transport in rat muscle cells and adipocytes. *Biochim Biophys Acta* 345: 387-404, 1974.

Clausen T and Hansen O. Active Na^+-K^+ transport and the rate of ouabain binding. The effect of insulin and other stimuli on skeletal muscle and adipocytes. *J Physiol (Lond)* 270: 415-430, 1977.

Clausen T, Hansen O, Kjeldsen K, and Nørgaard A. Effect of age, potassium depletion and denervation on specific displaceable [³H]ouabain binding in rat skeletal muscle in vivo. *J Physiol (Lond)* 333: 367-381, 1982.

Clausen T and Kohn PG. The effect of insulin on the transport of sodium and potassium in rat soleus muscle. *J Physiol (Lond)* 265: 19-42, 1977.

Clausen T and Nielsen OB. The Na^+,K^+-pump and muscle contractility. *Acta Physiol Scand* 152: 365-373, 1994.

Clausen T and Overgaard K. The role of K^+ channels in the force recovery elicited by Na^+-K^+ pump stimulation in Ba^{2+}-paralysed rat skeletal muscle. *J Physiol (Lond)* 527: 325-332, 2000.

Clausen T, Overgaard K, and Nielsen OB. Evidence that the Na^+ K^+ leak/pump ratio contributes to the difference in endurance between fast- and slow-twitch muscles. *Acta Physiol Scand,* 2004

Clausen T, Sellin LC, and Thesleff S. Quantitative changes in ouabain binding after denervation and during reinnervation of mouse skeletal muscle. *Acta Physiol Scand* 111: 373-375, 1981.

Clausen T, Van Hardeveld C, and Everts ME. Significance of cation transport in control of energy metabolism and thermogenesis. *Physiol Rev* 71: 733-774, 1991.

Clausen T, Wang P, Ørskov H, and Kristensen O. Hyperkalemic periodic paralysis. Relationships between changes in plasma water, electrolytes, insulin and catecholamines during attacks. *Scand J Clin Lab Invest* 40: 211-220, 1980.

Cox M, Sterns RH, and Singer I. The defence against hyperkalemia: the roles of insulin and aldosterone. *N Engl J Med* 299: 525-532, 1978.

Cougnon, MH, Mosely, AR, Radzyunkevich, TL, Lingrel, JB and Heiny, JA. Na,K-ATPase alpha and beta isoform expression in developing skeletal muscles alpha$_2$ correlates with t-tubule formation. *Pflügers Arch.* 445: 123-131, 2002.

Creese R. Sodium fluxes in diaphragm muscle and the effects of insulin and serum proteins. *J Physiol (Lond)* 197: 255-278, 1968.

DeClerck I, Boussery, K, Pannier, JL, and Van De Voorde J, Potassium potently relaxes small Rat Skeletal Muscle Arteries, *Medicine and Science in Sports and Exercise,* vol. 35: pp. 2005-2012, 2003.

DeFronzo RA, Sherwin RS, Dillingham M, Hendler R, Tamborlane WV, and Felig P. Influence of basal insulin and glucagon secretion on potassium and sodium metabolism. *J Clin Invest* 61: 472-479, 1978.

Desnuelle C, Lombet A, Serratrice G, and Lazdunski M. Sodium channel and sodium pump in normal and pathological muscles from patients with myotonic muscular dystrophy and lower motor neuron impairment. *J Clin Invest* 69: 358-367, 1982.

DiFranco, M, Hakimjavadi, H, Lingrel, JB and Heiny JA. Na^+,K^+-ATPase α_2 activity in mammalian skeletal muscle T-tubules is acutely stimulated by extracellular K^+. *J. Gen. Physiol.* 146(4):281-294, 2015.

Djurhuus MS, Vaag A, and Klitgaard NAH. Muscle sodium, potassium, and [^{3}H]ouabain binding in identical twins, discordant for type 2 diabetes. *J Clin Endocrinol Metab* 86: 859-866, 2001.

Dockry M, Kernan RP, and Tangney A. Active transport of sodium and potassium in mammalian skeletal muscle and its modification by nerve and by cholinergic and adrenergic agents. *J Physiol (Lond)* 186: 187-200, 1966.

Dombrowski L, Roy D, Marcotte B, and Marette A. A new procedure for the isolation of plasma membranes, t tubules, and internal membranes from skeletal muscle. *Am J Physiol Endocrinol Metab* 270: E667-E676, 1996.

Douglas RG, Breier BH, Gallaher BW, Koea JB, Shaw JH, and Gluckman PD. The circulating molecular weight forms of infused recombinant insulin-like growth factor-I and effects on glucose and fat metabolism in lambs. *Diabetologia*, 34(11):790-795, 1991.

Dua, AK, Dua, N, and Murrant, CL. Skeletal muscle contraction-induced vasodilator complement production is dependent on stimulus and contraction frequency. *Am. J. Physiol*, vol. 297:H433-H442, 2009.

Dutka, T.L. and Lamb, G.D. Na^+-K^+ pumps in the transverse tubular system of skeletal muscle fibers preferentially use ATP from glycolysis. *Am. J. Physiol.* vol. 293:C967-C977, 2007.

Dørup I. Magnesium and potassium deficiency, its diagnosis, occurence and treatment in diuretic therapy and its consequences for growth, protein synthesis and growth factors. Doctoral Thesis, *Acta Physiol Scand* 150 *Suppl* 618: 1-55, 1994.

Dørup I and Clausen T. ^{86}Rb is not a reliable tracer for potassium in skeletal muscle. *Biochem J* 302: 745-751, 1994.

Dørup I and Clausen T. Insulin-like growth factor I stimulates active Na^+-K^+ transport in rat soleus muscle. *Am J Physiol Endocrinol Metab* 268: E849-E857, 1995.

Dørup I and Clausen T. Effects of adrenal steroids on the concentration of Na^+-K^+ pumps in rat skeletal muscle. *J Endocrinol.* 152: 49-57, 1997.

Dørup I, Skajaa K, and Clausen T. A simple and rapid method for the determination of the concentrations of magnesium, sodium, potassium and sodium, potassium pumps in human skeletal muscle. *Clin Sci* 74: 241-248, 1988a.

Dørup I, Skajaa K, Clausen T, and Kjeldsen K. Reduced concentrations of potassium, magnesium, and sodium-potassium pumps in human skeletal muscle during treatment with diuretics. *Br Med J* 296: 455-458, 1988b.

Dørup I, Skajaa K, and Thybo NK. Oral magnesium supplementation restores the concentrations of magnesium, potassium and sodium-potassium pumps in skeletal muscle of patients receiving diuretic treatment. *J Int Med* 233: 117-123, 1993.

Dyken ML, Smith DM, and Peake RL. An electromyographic diagnostic screening test in McArdle's disease and a case report. *Neurology* 17: 45-50, 1967.

Edström L and Wroblewski R. Intracellular elemental composition of single muscle fibres in muscular dystrophy and dystrophia myotonica. *Acta Neurol Scand* 80: 419-424, 1989.

Erlij D and Grinstein S. Stimulation of the sodium pump by azide and high internal sodium: changes in the number of pumping sites and turnover rate. *J Physiol (Lond)* 259: 33-45, 1976.

Everts ME and Clausen T. Effects of thyroid hormone on Na^+-K^+ transport in resting and stimulated skeletal muscle. *Am J Physiol Endocrinol Metab* 255: E604-E612, 1988.

Everts ME and Clausen T. Excitation-induced activation of the Na^+ K^+ pump in rat skeletal muscle. *Am J Physiol Cell Physiol* 266: C925-C934, 1994.

Everts ME, Dørup I, Flyvbjerg A, Marshall SM, and Jørgensen KD. Na^+ K^+ pump in rat muscle: effects of hypophysectomy, growth hormone, and thyroid hormone. *Am J Physiol Endocrinol Metab* 259: E278-E283, 1990.

Everts ME, Skajaa K and Hansen K.The concentration of ouabain binding sites in biopsies of uterine muscle. *Acta Physiol. Scand.* 139: 503-510, 1990.

Evertsen F, Medbø JI, Jebens E, and Nicolaysen K. Hard training for 5 mo increases Na$^+$ K$^+$ pump concentration in skeletal muscle of cross-country skiers. *Am J Physiol Regul Integr Comp Physiol* 272: R1417-R1424, 1997.

Fenn WO. The role of potassium in physiological processes.*Physiol Rev* 20: 377-415, 1940.

Féraille E, Carranza ML, Rousselot M, and Favre H. Insulin enhances sodium sensitivity of Na$^+$ K$^+$ ATPase in isolated rat proximal convoluted tubule. *Am J Physiol Renal Fluid Electrolyte Physiol* 267: F55-F62, 1994.

Ferrannini E, Taddel S, Santoro D, Natali A, Boni C, Chiaro DD, and Buzzigoli G. Independent stimulation of glucose metabolism and Na$^+$-K$^+$ exchange by insulin in the human forearm. *Am J Physiol Endocrinol Metab* 255: E953-E958, 1988.

Flatman JA and Clausen T. Combined effects of adrenaline and insulin on active electrogenic Na$^+$ K$^+$ transport in rat soleus muscle. *Nature* 281: 580-581, 1979.

Fong CN, Atwood HL, and Charlton MP. Intracellular sodium activity at rest and after tetanic stimulation in muscles of normal and dystrophic (*dy2j/dy2j*)C57BL/6J mice. *Exp Neurol* 93: 359-368, 1986.

Ford GA, Dachman WD, BlaschTF, and Hoffman BB. Effect of aging on beta$_2$-adrenergic receptor-stimulated flux of K$^+$, PO$_4$, FFA, and glycerol in human forearms. *J Appl Physiol* 78: 172-178,1995.

Fredsted A., Gissel, H., Ørtenblad N. and T. Clausen. Effects of beta$_2$-agonists on force during and following anoxia in rat extensor digitorum longus muscle. *J. Appl. Physiol* 112: 2057-2067, 2012.

Fuller W, Eaton P, Bell, JR and Shattock, MJ Ischemia-induced phosphorylation of phospholemman directly activates rat cardiac Na/K-ATPase *The FASEB Journal* vol. 18: 197-199, 2004.

Galbo H, Holst JJ, and Christensen NJ. Glucagon and plasma catecholamine responses to graded and prolonged exercise in man. *J Appl Physiol* 38: 70-76, 1975.

Giordano M and DeFronzo RA. Acute effect of human recombinant insulin-like growth factor I on renal function in humans. *Nephron* 71: 10-15, 1995.

Glitsch HG and Tappe A. The Na$^+$/K$^+$ pump of cardiac Purkinje cells in preferentially fuelled by glycolytic ATP production. *Pflügers Arch* 422: 380-385, 1993.

Glynn I. A Hundred Years of Sodium Pumping. *Annu. Rev. Physiol.* 64:1-18, 2002.

Goodman CA, Bennie JA, Leikis MJ and McKenna MJ. Unaccustomed Eccentric Contractions Impair Plasma K$^+$ Regulation in the Absence of Changes in Muscle Na$^+$,K$^+$-ATPase Content. *PLoS One.* 9(6): e101039, 2014

Green HJ, Burnett ME, D'Arsigny CL, Webb KA, McBride I, Ouyang J, and O'Donnell Vastus Lateralis Na$^+$-K$^+$-ATPase Activity, protein, and isoform distribution in chronic obstructive pulmonary disease *Muscle Nerve* 40:62-68, 2009.

Green HJ, Chin ER, Ball-Burnett M, and Ranney D. Increases in human skeletal muscle Na$^+$-K$^+$-ATPase concentration with short-term training. *Am J Physiol Cell Physiol* 264: C1538-C11541,1993.

Green HJ, MacDougall J, Tarnopolsky M, and Melissa NL. Downregulation of Na$^+$-K$^+$-ATPase in skeletal muscle with training in normobaric hypoxia. *J Appl Physiol* 86: 1745-1748, 1999.

Green HJ, Duscha BD, Sullivan MJ, Keteyian SJ, and Kraus WE. Normal skeletal muscle Na$^+$-K$^+$pump concentration in patients with chronic heart failure. *Muscle Nerve* 24: 69-76, 2001.

Green HJ, Roy B, Grant S, Burnett M, Tupling R, Otto C, Pipe A, and McKenzie D. Downregulation in muscle Na$^+$-K$^+$ ATPase following a 21-day expedition to 6,194 m. *J Appl Physiol* 88: 634-640, 2000a.

Green S, Langberg H, Skovgaard D, Bülow J, and Kjaer M. Interstitial and arterial-venous

[K$^+$] in human calf muscle during dynamic exercise: effect of ischaemia and relation to muscle pain. *J Physiol (Lond)* 529: 849-861, 2000b.

Green HJ, Burnett ME, D'Arsigny CL, Webb KA, McBride I, Ouyang J, and O'Donnell Vastus Lateralis Na$^+$-K$^+$-ATPase Activity, protein, and isoform distribution in chronic obstructive pulmonary disease *Muscle Nerve* 40:62-68, 2009.

Griggs RC, Mendell JR, and Miller RG. *Evaluation and Treatment of Myopathies.* Philadelphia, PA: Davis, 1995.

Gruener R, Stern LZ, Markovitz D, and Gerdes C. Electrophysiologic properties of intercostal muscle fibers in human neuromuscular diseases. *Muscle Nerve* 2: 165-172, 1979.

Gullestad L, Hallen J, and Sejersted OM. K_ balance of the quadriceps muscle during dynamic exercise with and without beta-adrenoceptor blockade. *J Appl Physiol* 78:513-523, 1995.

Haller RG, Clausen T and Vissing J. Reduced levels of skeletal muscle Na$^+$,K$^+$-ATPase in McArdle disease. *Neurology* 50:37-40, 1998.

Hansen O and Clausen T. Quantitative determination of Na$^+$-K$^+$- ATPase and other sarcolemmal components in muscle cells. Editorial review. *Am J Physiol Cell Physiol* 254: C1-C7,1988.

Hansen O, Jensen J, Nørby JG, and Ottolenghi P. A new proposal regarding the subunit composition of (Na$^+$,K$^+$)ATPase. *Nature* 280: 410-412, 1979.

Hansen PS, Buhagiar KA, Gray DF, and Rasmussen HH. Voltage- dependent stimulation of the Na$^+$K$^+$ pump by insulin in rabbit cardiac myocytes. *Am J Physiol Cell Physiol* 278: C546-C553, 2000.

Harrison AP and Clausen T. Thyroid hormone-induced upregulation of Na$^+$ channels and Na$^+$-K$^+$ pumps: implications for contractility. *Am J Physiol Regul Integr Comp Physiol* 274: R864-R867, 1998.

Harrison AP, Clausen T, Duchamp C, and Dauncey MJ. Roles of skeletal muscle morphology and activity in determining Na$^+$- K$^+$- ATPase concentration in young pigs. *Am J Physiol Regul Integr Comp Physiol* 266: R102-R111, 1994.

Harrison AP, Nielsen OB, and Clausen T. Role of Na$^+$-K$^+$ pump and Na$^+$ channel concentrations in the contractility of rat soleus muscle. *Am J Physiol* 272: R1402-R1408, 1997.

Hays, ET, Dwyer, TM, Horowicz, P and Swift, JG.: Epinephrine action on sodium fluxes in frog striated muscle. *Am. J. Physiol.* 227, 1340-1347, 1974.

Hodgkin AL and Horowicz P. Movements of Na and K in single muscle fibres. *J Physiol (Lond)* 145: 405-432, 1959.

Homonko DA and Theriault E. Downhill running preferentially increases CGRP in fast glycolytic muscle fibers. *J Appl Physiol* 89: 1928-1936, 2000.

Hostrup M, Jessen S, Onslev J, Clausen T, Porsbjerg CM. Two-week inhalation of budesonide increases muscle Na$^+$ K$^+$-ATPase content but not endurance in response to terbutaline in man. *Scand. J. Med. Sci. Sports 2016.*

Hsu YM and Guidotti G. Effects of hypokalemia on the properties and expression of the (Na$^+$, K$^+$)-ATPase of rat skeletal muscle *J Biol Chem* 266: 427-433, 1991.

Hundal HS, Marette A, Mitsumoto Y, Ramlal T, Blostein R, and Klip A. Insulin induces translocation of the α_2 and $\beta 1$ subunits of the Na$^+$,K$^+$-ATPase from intracellular compartments to the plasma membrane in mammalian skeletal muscle. *J Biol Chem* 267:5040-5043, 1992.

Iannaconne ST, Li KX, Sperelakis N, and Lathrop DA. Insulin-induced hyperpolarization in mammalian skeletal muscle. *Am J Physiol Cell Physiol* 256: C368-C374, 1989.

James JH, Wagner KR, King JK, Leffler RE, Upputuri RK, Balasubramaniam A, Friend LA, Shelly DA, Paul RJ, and Fischer JE. Stimulation of both aerobic glycolysis and Na$^+$ K$^+$- ATPase activity in skeletal muscle by epinephrine or amylin. *Am J Physiol Endocrinol Metab* 277: E176-E186, 1999.

Jebens E, Steen H, Fjeld TO, Bye E, and Sejersted OM. Changes in Na+,K+-adenosine-triphosphatase, citrate synthase and K+ in sheep skeletal muscle during immobilization and remobilization. *Eur J Appl Physiol* 71: 386-395, 1995.

Jones DA. Muscle fatigue due to changes beyond the neuromuscular junction. *Ciba Found Symp* 82: 178-196, 1981.

Juel C. The effect of beta2-adrenoceptor activation on ion-shifts and fatigue in mouse soleus muscles stimulated in vitro. *Acta Physiologica* (Print Edition), Vol. 134, p. 209-216, 1988.

Kanbe M and Kitasato H. Stimulation of Na,K-ATPase activity of frog skeletal muscle by insulin. *Biochem Biophys Res Commun* 134: 609-616, 1986.

Kashihara Y, Sakaguchi M, and Kuno M. Axonal transport and distribution of endogenous calcitonin gene-related peptide in rat peripheral nerve. *J Neurosci* 9: 3796-3802, 1989.

Kirsch GE, Nichols RA, and Nakajima S. Delayed rectification in the transverse tubules. Origin of the late after-potential in the frog skeletal muscle. *J Gen Physiol* 70: 1-21, 1977.

Kitasato H, Sato S, Marunaka Y, Murayama K, and Nishio K. Apparent affinity changes induced by insulin of Na-K transport system in frog skeletal muscle. *Jpn J Physiol* 30: 603-616, 1980.

Kjeldsen K. Complete quantification of the total concentration of rat skeletal-muscle Na+ K+-dependent ATPase by measurements of [3H]ouabain binding. *Biochem J* 240: 725-730, 1986.

Kjeldsen K. The importance of skeletal muscles as a distribution volume for digitalis glycosides in guinea-pigs and human subjects. In: *Cardiac Glycosides*, edited by Erdmann E, Greeff K, and Skou C. Darmstadt, Germany: Steinkopff Verlag, 1986, pp. 245-255.

Kjeldsen K. and Nørgaard A. Quantification of rat sciatic nerve Na, K+-ATPase by measurements of 3H-ouabain binding in intact nerve samples. *J. Neurol. Sci.*, 79, 205-219. 1987

Kjeldsen K and Grøn P. Skeletal muscle Na+,K+-pump concentration in children and its relationship to cardiac glycoside distribution. *J Pharmacol Exp Ther* 250: 721-725, 1989.

Kjeldsen K, Everts ME, and Clausen T. The effects of thyroid hormones on [3H]ouabain binding site concentration, Na+,K+-contents and 86Rb-efflux in rat skeletal muscle. *Pflügers Arch* 406: 529-535, 1986a.

Kjeldsen K, Everts ME, and Clausen T. Effects of semi-starved and potassium deficiency on the concentration of [3H]ouabain binding sites and sodium and potassium contents in rat skeletal muscle. *Br J Nutr* 56: 519-532, 1986b.

Kjeldsen K, Richter EA, Galbo H, Lortie G, and Clausen T. Training increases the concentration of 3H-ouabain binding sites in rat skeletal muscle. *Biochim. Biophys. Acta* 860: 708-727, 1986c.

Kjeldsen K, Gøtzsche CO, Nørgaard A, Thomassen A, and Clausen T. Effect of thyroid function on number of Na-K pumps in human skeletal muscle. *Lancet* ii: 8-10, 1984.

Kjeldsen K, Nørgaard A, and Clausen T. The age-dependent changes in the number of [3H]ouabain binding sites in mammalian skeletal muscle. *Pflügers Arch* 402: 100-108, 1984.

Kjeldsen K, Nørgaard A, and Clausen T. Effect of K+-depletion on [3H]ouabain binding and Na+, K+-contents in mammalian skeletal muscle. *Acta Physiol Scand* 122: 103-117, 1984c.

Kjeldsen K, Brændgaard H, Sidenius P, Larsen JS and Nørgaard A. Diabetes decreases Na+, K+ pump concentration in skeletal muscles, heart ventricular muscle and peripheral nerves of rat. *Diabetes* 36: 842-848, 1987.

Klitgaard H and Clausen T. Increased total concentration of Na+, K+ pumps in vastus lateralis muscle of old trained human subjects. *J Appl Physiol* 67: 2491-2494, 1989.

Knochel JP, Blachley JD, JohJH, and Carter NW. Muscle cell electrical hyperpolarization and reduced exercise hyperkalemia in physically conditioned dogs. *J Clin Invest* 75: 740-745, 1985.

Lavoie L, Roy D, Ramlal T, Dombrowski L, Martin-Vasallo P, Marette A, Carpentier JL, and Klip A. Insulin-induced translocation of Na$^+$-K$^+$-ATPase subunits to the plasma membrane is muscle fiber type specific. *Am J Physiol Cell Physiol* 270: C1421- C1429, 1996.

Lehmann-Horn F and Jurkat-Rott K. Voltage-Gated Ion Channels and Hereditary Disease. *Physiological Reviews, Vol. 79 no. 4, 1317-1372, 1999.*

Leivseth G and Reikeras O. Changes in muscle fiber cross sectional area and concentrations of Na$^+$, K$^+$-ATPase in deltoid muscle in patients with impingement syndrome of the shoulder. *J Orthop Sports Phys Ther* 19:146-149, 1994.

Li KX and Sperelakis N. Isoproterenol- and insulin-induced hyperpolarization in rat skeletal muscle. *J Cell Physiol* 157: 631-636, 1993.

Li KX and Sperelakis N. Electrogenic Na-K pump current in rat skeletal myoballs. *J Cell Physiol* 159: 181-186, 1994.

Lingrel, J.B. The physiological significance of the cardiotonic steroid/ouabain-binding site of the Na,K-ATPase. *Ann.Rev. Physiol.* 72. 395-412, 2010

Lippe C, Bellantuono V, Castronuovo G, Ardizzone C, and Cassano G. Action of capsaicin and related peptides on the ionic transport across the skin of *Rana esculenta*. *Arch Int Physiol Biochim Biophys* 102: 51-54, 1994.

Lo A. and Fuglevand, AJ., and Secomb. Theoretical simulation of K$^+$-based mechanisms for regulation of capillary perfusion in skeletal muscle. *Am. J. Physiol.* vol. 287: H833-H840, 2004.

Longo N, Scaglia F, and Wang Y. Insulin increases the turnover rate of Na$^+$ K$^+$-ATPase in human fibroblasts. *Am J Physiol Cell Physiol* 280: C912-C919, 2001.

Lytton J. Insulin affects the sodium affinity of the rat adipocyte (Na$^+$, K$^+$)-ATPase. *J Biol Chem* 260: 10075-10080, 1985.

Macdonald, W.A., Nielsen O.B. and Clausen, T. Na$^+$, K$^+$ pump stimulation restores carbacholine-induced loss of excitability and contractility in rat skeletal muscle, *J. Physiol.* 563: 459-469, 2005.

Madsen K, Franch J, and Clausen T. Effects of intensified endurance training on the concentration of Na$^+$, K$^+$-ATPase and Ca-ATPase in human skeletal muscle. *Acta Physiol Scand* 150: 251-258, 1994.

Marcos E and Ribas J. Kinetics of plasma potassium concentrations during exhausting exercise in trained and untrained men. *Eur J Appl Physiol* 71: 207-214, 1995.

Marette A, Krischer J, Lavoie L, Ackerley C, Carpentier JL, and Klip A. Insulin increases the Na$^+$, K$^+$-ATPase α_2-subunit in the surface of rat skeletal muscle: morphological evidence. *Am J Physiol Cell Physiol* 265: C1716-C1722, 1993.

Matsumura M, Kuzuya N, Kawakami Y, and Yamashita K. Effects of fasting, refeeding, and fasting with T3 administration on Na$^+$-K$^+$, ATPase in rat skeletal muscle. *Metabolism* 41: 995-999, 1992.

McCarter FD, James JH, Luchette FA, Wang L, Friend LA, King JK, Evans JM, George MA, and Fischer JE. Adrenergic blockade reduces skeletal muscle glycolysis and Na$^+$, K$^+$-ATPase activity during hemorrhage. *J Surg Res* 99: 235-244, 2001.

McKenna MJ, Gissel H, and Clausen T. Effects of electrical stimulation and insulin on Na$^+$, K$^+$-ATPase (^{3}H-ouabain binding) in rat skeletal muscle. *J Physiol (Lond)* 547: 567-580, 2003.

McKenna MJ, Schmidt TA, Hargreaves M, Cameron L, Skinner SL, and Kjeldsen K. Sprint training increases human skeletal muscle Na$^+$, K$^+$-ATPase concentration and improves K$^+$ -regulation. *J Appl Physiol* 75: 173-180, 1993.

McKenna MJ, Perry BD, Serpiello FR, Caldow MK, Levinger P, Cameron-Smith

D, Levinger I. Unchanged [^{3}H]ouabain binding site content but reduced Na$^+$, K$^+$ pump α2-protein abundance in skeletal muscle in older adults. J Appl Physiol., Vol. 113(10):1505-1511, 2012.

McMorrow C, Fredsted A, Carberry J, O'Connell RA, Bradford A, Jones JF, O'Halloran KD. Chronic hypoxia increases rat diaphragm muscle endurance and sodium-potassium ATPase pump content. Eur Respir J. 37(6):1474-81, 2011.

Medbø, J. I. and Sejersted, O.M. Plasma potassium changes with high intensity exercise. *J Physiol.* 421: 105-122, 1990.

Mikkelsen, U.R. Gissel, H., Fredsted, A. and Clausen, T. Excitation-induced cell damage and β_2-adrenoceptor agonist stimulated force recovery in rat skeletal muscle *Am. J. Physiol.* 290:R265-R272, 2006.

Moore RD. Effect of insulin upon the sodium pump in frog skeletal muscle. *J Physiol (Lond)* 232: 23-45, 1973.

Moore RD, Munford JW, and Pillsworth TJ Jr. Effects of streptozotocin diabetes and fasting on intracellular sodium and adenosine triphosphate in rat soleus muscle. *J Physiol (Lond)* 338: 277-294, 1983.

Moore LE and Tsai TD. Ion conductances of the surface and transverse tubular membranes of skeletal muscle. *J Membr Biol* 73: 217-226, 1983.

Müller-Ehmsen J, Juvvadi P, Thompson CB, Tumyan L, Croyle M, Lingrel JB, Schwinger RHG, McDonough AA, and Farley RA. Ouabain and substrate affinities of human Na$^+$, K$^+$-ATPase alpha(1)beta(1), alpha(2)beta(1), and alpha(3)beta(1) when expressed separately in yeast cells. *Am J Physiol Cell Physiol* 281: C1355-C1364, 2001.

Mu X, Peng H, Pan H, Huard J, Li Y. Study of muscle cell dedifferentiation after skeletal muscle injury of mice with a Cre-Lox system. *PLoS One*, Feb 3;6(2):e16699, 2011.

Murphy, KT, Nielsen, OB and Clausen, T. Analysis of exercise-induced Na$^+$, K$^+$ exchange in rat skeletal muscle *in vivo*. *Exp. Phys.* Volume 93, Issue 12, Pages 1249-1262, 2008.

Natali A, Quinones GA, Santoro D, Pecori N, Taddei S, Salvetti A, and Ferrannini E. Relationship between insulin release, antinatriuresis and hypokalaemia after glucose ingestion in normal and hypertensive man. *Clin Sci* 85: 327-335, 1993.

Ng YC, Tolerico PH, and Book CBS. Alterations in the levels of Na$^+$, K$^+$-ATPase isoforms in heart, skeletal muscle and kidney of diabetic rats. *Am J Physiol Endocrinol Metab* 265: E243-E251, 1993.

Nielsen OB and Clausen T. The significance of active Na$^+$, K$^+$ transport in the maintenance of contractility in rat skeletal muscle. *Acta Physiol Scand.* 157: 199-209, 1996.

Nielsen OB and Clausen T. Regulation of Na$^+$-K$^+$ pump activity in contracting rat muscle. *J Physiol (Lond)* 503: 571-581, 1997.

Nielsen OB & Clausen T. The Na$^+$, K$^+$-pump protects muscle excitability and contractility during exercise, *Exercise and Sport Sciences Reviews*, vol. 28, pp. 159-164, 2000.

Nielsen OB and Harrison AP. The regulation of the Na$^+$, K$^+$ pump in contracting skeletal muscle, invited review. *Acta PhysiolScand* 162: 191-200, 1998.

Nielsen OB, Hilsted L, and Clausen T. Excitation-induced force recovery in potassium-inhibited rat soleus muscle. *J Physiol (Lond)* 512: 819-829, 1998.

Nishida K, Ohara T, JohJ, Wallner JS, Wilk J, Sherman N, Kawakami K, Sussman KE, and Draznin B. Na$^+$, K$^+$-ATPase activity and its αII subunit gene expression in rat skeletal muscle: influence of diabetes, fasting, and refeeding. *Metabolism* 41: 56-63, 1992.

Nordsborg, N, Mohr, M, Pedersen, LD, Nielsen, JJ, Langberg, H, Bangsbo, J. Muscle interstitial potassium kinetics during intense exhaustive exercise: effect of previous arm exercise. *American Journal of Physiology – Regulatory, Integrative and Comparative Physiology*, Vol. 285 no. 1, R143-R148, 2003

Nordsborg, N. Goodmann, C, McKenna, M. J. and Bangsbo. J. Dexamethasone up-regulates skeletal muscle maximal Na$^+$, K$^+$ pump activity by muscle group specific mechanisms in humans. *J. Physiol.* 567: 583-589, 2005.

Nordsborg, N, Ovesen, J, Thomassen, M, Zangenberg, M, Jøns, C, Iaia, FM, Nielsen, JJ, Bangsbo, J. Effect of dexamethasone on skeletal muscle Na^+,K^+ pump subunit specific expression and K^+ homeostasis during exercise in humans. *J. of Physiol*, Volume 586, Issue 5, Pages 1447-1459, 2008

Northcote, R. J. Flannigan, C. and Ballantyne, D. Sudden death and vigorous exercise – a study of 60 deaths associated with squash. *Br. Heart J.* 55: 198-203, 1986.

Nørgaard A, Bagger JP, Bjerregaard, P, Baandrup U, Kjeldsen K, and Thomsen PEB. Relation of left ventricular function and Na^+,K^+-pump concentration in suspected idiopathic dilated cardiomyopathy. *Am J Cardiol* 61: 1312-1315, 1988.

Nørgaard A, Jensen JH, and Andreasen F. Effect of amiodarone on ^{3}H-ouabain binding sites in human skeletal muscle. *Eur J Clin Pharmacol* 38: 397-399, 1990.

Nørgaard A, Kjeldsen K, and Clausen T. Potassium depletion decreases the number of ^{3}H-ouabain binding sites and the active Na^+,K^+ transport in skeletal muscle. *Nature* 293: 739-741, 1981.

Nørgaard A, Kjeldsen K, and Clausen T. A method for the determination of the total number of ^{3}H-ouabain binding sites in biopsies of human skeletal muscle. *Scand J Clin Lab Invest* 44: 509-518, 1984.

Nørgaard A, Kjeldsen K, and Hansen O. (Na^+ K^+)-ATPase activity of crude homogenates of rat skeletal muscle as estimated from their K^+-dependent 3-*O*-methylfluorescein phosphatase activity. *Biochim Biophys Acta* 770: 203-209, 1984.

Nørgaard A, Kjeldsen K, and Hansen O. K^+-dependent 3-*O*-methylfluorescein phosphatase activity in crude homogeneate of rodent heart ventricle: effect of K^+ depletion and changes in thyroid status. *Eur J Pharmacol* 113: 373-382, 1985.

Nørgaard A, Kjeldsen K, Hansen O, and Clausen T. A simple and rapid method for the determination of the number of ^{3}Houabain binding sites in biopsies of skeletal muscle. *Biochem Biophys Res Commun* 111: 319-325, 1983.

Ogawa, H. Shinoda, T. Cornelius F. and Toyoshima C. *Crystal structure of the sodium-potassium pump (Na^+, K^+ ATPase) with bound potassium and ouabain. Proc. Natl. Acad. Sci, USA, 106,13742-13747, 2009.*

Omatsu-Kanbe M and Kitasato H. Effects of detergents on $Na^+·K^+$-dependent ATPase activity in plasma-membrane fractions prepared from frog muscles. *Biochem J* 246: 583-588, 1987.

Omatsu-Kanbe M and Kitasato H. Insulin stimulates the translocation of Na^+,K^+-dependent ATPase molecules from intracellular stores to the plasma membrane in frog skeletal muscle. *Biochem J* 272: 727-733, 1990.

Onuoha GN, Nicholls DP, Patterson A, and Beringer T. Neuropeptide secretion in exercise. *Neuropeptides* 32: 319-325, 1998.

Orlowski J and Lingrel JB. Tissue-specific and developmental regulation of rat Na^+,K^+-ATPase catalytic α isoform and β subunit mRNAs. *J Biol Chem* 263: 10436-10442, 1988.

Overgaard K and Nielsen OB. Activity-induced recovery of excitability in K^+-depressed rat soleus muscle. *Am J Physiol Regul Integr Comp Physiol* 280: R48-R55, 2001.

Overgaard K, Nielsen OB, and Clausen T. Effects of reduced electrochemical Na^+ gradient on contractility in skeletal muscle: role of the $Na^+·K^+$-pump. *Pflügers Arch* 434:457-465, 1997.

Overgaard K, Nielsen OB, Flatman JA, and Clausen T. Relations between excitability and contractility in rat soleus muscle: role of the $Na^+·K^+$-pump and $Na^+·K^+$ gradients. *J Physiol (Lond)* 518: 215-225, 1999.

Paterson DJ, Friedland JS, Bascom DA, Clement ID, Cunningham DA, Painter R, and Robbins PA. Changes in arterial K^+ and ventilation during exercise in normal subjects and subjects with McArdle's syndrome. *J Physiol (Lond)* 429: 339-348, 1990.

Petersen AC, Leikis MJ, McMahon LP, Kent AB, Murphy KT, Gong X, and McKenna MJ. Impaired exercise performance and muscle Na+,K+-pump activity in renal transplantation and haemodialysis patients. *Nephrol Dial Transplant*, 27(5): 2036-2043, 2012.

Peachey LD and Eisenberg BR. Helicoids in the T system and striations of frog skeletal muscle fibers seen by high voltage electron microscopy. *Biophys J* 22: 145-154, 1978.

Pedersen TH, Clausen T, and Nielsen OB. Loss of force induced induced by high extracellular [K+] in rat muscle: effect of temperature, lactic acid amd β2-agonist. *J. Physiol.* Vol. 551: pp. 277-286, 2003.

Perry BD, Levinger P, Morris HG, Petersen AC, Garnham AP, Levinger I, and McKenna MJ. The effects of knee injury on skeletal muscle function, Na+,K+-ATPase content, and isoform abundance. *Physiol Rep.*, Feb. 2015.

Pollack LR, Tate EH, and Cook JS. Na+,K+-ATPase in HeLa cells after prolonged growth in low K+ or ouabain. *J Cell Physiol* 106: 85-97, 1981.

Powell WJ and Skinner NS. Effect of the catecholamines on ionic balance and vascular resistance in skeletal muscle. *Am J Cardiol* 18: 73 -82, 1966.

Ragolia L, Cherpalis B, Srinivasan M, and Begum N. Role of serine/threonine protein phosphatases in insulin regulation of Na+,K+-ATPase activity in cultured rat skeletal muscle cells. *J Biol Chem* 272: 23653-23658, 1997.

Ramlal T, Ewart HS, Somwar R, Deems RO, Valentin MA, Young DA, and Klip A. Muscle subcellular localization and recruitment by insulin of glucose transporters and Na+,K+-ATPase subunits in transgenic mice overexpressing the GLUT4 glucose transporter. *Diabetes* 45: 1516-1523, 1996.

Ravn HB and Dørup I. The concentration of sodium,potassium pumps in chronic obstructive lung disease (COLD) patients: the impact of magnesium depletion and steroid treatment. *J Intern Med* 241: 23-29, 1997.

Resh MD, Nemetoff RA, and Guidotti G. Insulin stimulation of Na+,K+-adenosine triphosphatase-dependent 86Rb- uptake in rat adipocytes. *J Biol Chem* 255: 10938-10945, 1980.

Riis ALD, Jørgensen, JOL, Møller, N, Weeke and Clausen, T. Hyperthyroidism and cation pumps in human skeletal muscle. *Am. J. Physiol.* 288: E1265-1269, 2005.

Richter EA, Sonne B, Christensen NJ, and Galbo H. Role of epinephrine for muscular glycogenolysis and pancreatic hormonal secretion in running rats. *Am J Physiol Endocrinol Metab* 240: E526-E532,1981.

Robertson D, Frölich JC, Carr RK, Watson JT, Hollifield JW, Shand DG, and Oates JA Effects of caffeine on plasma renin activity, catecholamines and blood pressure. *N. Engl. J. of Medicine*, Vol. 298: pp. 181-186, 1978.

Sacco P, McIntyre DB, and Jones DA. Effects of length and stimulation frequency on fatigue of the human tibialis anterior muscle. *J Appl Physiol* 77: 1148-1154, 1994.

Sadre M, Sheng HP, Fiorotto M, and Nichols BL. Electrolyte composition changes of chronically K+-depleted rats after K+ loading. *J Appl Physiol* 63: 765-769, 1987.

Sakaguchi M, Inaishi Y, Kashihara Y, and Kuno M. Release of calcitonin gene-related peptide from nerve terminals in rat skeletal muscle. *J Physiol (Lond)* 434: 257-270, 1991.

Sampson SR, Brodie C, and Alboim SV. Role of protein kinase C in insulin activation of the Na-K pump in cultured skeletal muscle. *Am J Physiol Cell Physiol* 266: C751-C758, 1994.

Santicioli P, Del Bianco E, Geppetti P, and Maggi CA. Release of calcitonin gene-related peptide-like (CGRP-LI) immunoreactivity from rat isolated soleus muscle by low pH, capsaicin and potassium. *Neurosci Lett* 143: 19-22, 1992.

Schmidt TA, Holm-Nielsen P, and Kjeldsen K. Human skeletal muscle digitalis glycoside receptors (Na+,K+-ATPase): importance during digitalization. *Cardiovasc Drugs Ther* 7: 175-181, 1993.

Schmidt TA, Hasselbalch S, Farrell PA, Vestergaard H, and Kjeldsen K. Human and rodent muscle Na⁺,K⁺-ATPase in diabetes related to insulin, starvation, and training. *J Appl Physiol* 76: 2140-2146, 1994.

Schmidt TA, Svendsen JH, Haunsø S, and Kjeldsen K. Quantification of the total Na⁺,K⁺-ATPase concentration in atria and ventricles from mammalian species by measuring ³H-ouabain binding to intact myocardial samples. Stability to short term ischemia reperfusion. *Basic Res Cardiol* 85: 411-427, 1990.

Schmidt TA, Larsen JS, and Kjeldsen K. Quantification of rat cerebral cortex Na⁺·K⁺-ATPase: Effect of age and potassium depletion, *J. of Neurochemistry* 59:2094-2104 1992. (Sect. 10/1)

Schmidt TA, Hasselbalch S, Larsen JS, Bundgaard H, Juhler M, and Kjeldsen K. Reduction of cerebral cortical ³Houabain binding site (Na⁺·K⁺-ATPase) density in dementia as evaluated in fresh human cerebral cortical biopsies. *Cognitive Brain Research* 4:281-287, 1996.

Schmitz O, Alberti KGMM, Hreidarsson AB, Laurberg P, and Weeke J. Suppression of the night increase in serum TSH during development of ketosis in diabetic patients. *J Endocrinol Invest* 4: 403-407, 1981.

Schussler GC and Orlando J. Fasting decreases triiodothyronine receptor capacity. *Science* 199: 686-688, 1978.

Sejersted OM and Sjøgaard G. Dynamics and consequences of potassium shifts in skeletal muscle and heart during exercise. *Physiol Rev* 80: 1411-1481, 2000.

Shattock, M.J. Phospholemman: its role in normal cardiac physiology and potential as a drugable target in disease. *Curr. Opin. Pharmacol.* 9: 160-166, 2009.

Shinoda, T., Ogawa, H., Cornelius, F. and Toyoshima, C. Crystal structure of the sodium-potassium pump at 2.4 A resolution. Nature. May 21;459(7245):446-50, 2009.

Simonides, W.S. and Van Hardeveld, C. (Ca²⁺-Mg²⁺)-ATPase activity associated with the maintenance of a Ca²⁺ gradient by sarcoplasmic reticulum at submicromolar ex-

ternal [Ca²]. The effect of hypothyroidism. *Biochim. Biophys. Acta*, 943:349-359, 1988.

Sjøgaard G. Exercise-induced muscle fatigue: the significance of potassium. *Acta Physiol Scand Suppl* 593: 1-63, 1990.

Skou JC. Enzymatic basis for active transport of Na⁺ and K⁺ across cell membrane. *Physiol Rev* 45: 596-617, 1965.

Skou JC. The Na-K pump. *News in Physiological Sciences*, 7:95-100, 1992

Skov M, Vincenzo de Paoli F, Lausten J, Bækgaard Nielsen O, Holm Pedersen T. Extracellular magnesium and calcium reduce myotonia in isolated CLC-1 chloride channel-inhibited human muscle. *Muscle Nerve*, 51:65-71, 2015.

Somlyo AP and Somlyo AV. Pharmacology of excitation-contraction coupling in vascular smooth muscle and in avian slow muscle. *Federation Proc* 28: 1634-1642, 1969.

Sowinski, K. Cronin, D. Mueller, B. A. and Kraus, M. A. Subcutaneous terbutaline use in CKD to reduce Potassium Concentrations. *Am. J. Kidney Dis.* 45:1040-1045, 2005.

Sundaresan PR, Sharma VK, Gingold SI, and Banerjee SP. Decreased β-adrenergic receptors in rat heart in streptozotocin-induced diabetes: role of thyroid hormones. *Endocrinology* 114:1358-1363, 1984.

Suwannachot P, Verkleij CB, Kocsis S, Enzerink E, and Everts ME. Prolonged food restriction and mild exercise in Shetland ponies: effects on weight gain, thyroid hormone concentrations and muscle Na⁺,K⁺-ATPase. *J Endocrinol* 167: 321-329, 2000.

Suwannachot P, Verkleij CB, Kocsis S, van Weeren PR, and Everts ME. Specificity and reversibility of the training effects on the concentration of Na⁺,K⁺-ATPase in foal skeletal muscle. *Equine Vet J* 33: 250-255, 2001.

Swann AC. Caloric intake and Na⁺-K⁺-ATPase: differential regulation by alpha₁- and beta-noradrenergic receptors. *Am J Physiol Regul Integr Comp Physiol* 247: R449-R455, 1984.

Takami K, Hashimoto K, Uchida S, Tohyama M, and Yoshida H. Effect of calcitonin gene-related peptide on the cyclic AMP level of isolated mouse diaphragm. *Jpn J Pharmacol* 42: 345-350, 1986.

Tashiro N. Effects of isoprenaline on contractions of directly stimulated fast and slow skeletal muscles of the guinea-pig. *Br J Pharmacol* 48: 121-131, 1973.

Vagenakis AG, Burger A, Portnary GI, Rudolph M, O'Brian Azizi F Jr, Arky RA, Nicod P, Ingbar SH, and Braverman LE. Diversion of peripheral thyroxine metabolism from activating to inactivating pathways during complete fasting. *J Clin Endocrinol Metab* 41: 191-194, 1975.

Van Hardeveld, C. and Clausen, T. Effect of thyroid status on K$^+$-stimulated metabolism and ^{45}Ca-exchange in rat skeletal muscle. *Am. J. Physiol.* 247:E421-E430, 1984.

Vine W, Smith P, LaChappell R, Blase E, and Young A. Effects of rat amylin on renal function in the rat. *Horm Metab Res* 30: 518-522, 1998.

Voldstedlund M, Tranum-Jensen J, and Vinten J. Quantitation of Na$^+$/K$^+$ATPase and glucose transporter isoforms in rat adipocyte plasma membrane by immunogold labeling. *J Membr Biol* 136: 63-73, 1993.

Wallinga W, Meijer SL, Alberink MJ, Vlick M, Wienk ED, and Ypey DL. Modelling action potentials and membrane currents of mammalian skeletal muscle fibres in coherence with potassium concentration changes in the T-tubular system. *Eur Biophys J* 28: 317-329, 1999.

Wang J, Velotta JB, McDonough AA, and Farley RA. All human Na$^+$-K$^+$-ATPase -subunit isoforms have a similar affinity for cardiac glycosides. *Am J Physiol Cell Physiol* 281: C1336-C1343, 2001.

Wang P and Clausen T. Treatment of attacks in hyperkalemic familial periodic paralysis by inhalation of salbutamol. *Lancet* i: 221, 1976.

Ward KM, Manning W, and Wareham AC. Effects of denervation and immobiliation during development upon [^{3}H]ouabain binding by slow- and fast-twitch muscle of the rat. *J Neurol Sci* 78: 213-224, 1987.

Ward KM and Wareham AC. Intracellular activity of sodium in normal and dystrophic skeletal muscle from C57BL/6J mice. *Exp Neurol* 83: 629-633, 1984.

Weil E, Sasson S, and Gutman Y. Mechanisms of insulin-induced activation of Na$^+$-K$^+$-ATPase in isolated rat soleus muscle. *Am J Physiol Cell Physiol* 261: C224-C230, 1991.

Wevers RA, Joosten MG, van de Biezenbos JB, Theewes GM, and Veerkamp JH. Excessive plasma K$^+$ increase after ischemic exercise in myotonic muscular dystrophy. *Muscle Nerve* 13: 27-32, 1990.

Wijkerslooth, LRH de, Koch, BCP, Malingré, MM, Smits, P and Bartelink, AKM. Life-threatening hypokalaemia and lactate accumulation after autointoxication with Stacker 2$^®$, a 'powerful slimming agent'. *British Journal of Clinical Pharmacology* Volume 66, Issue 5, pages 728-731, 2008.

Williams MRW, Resneck WG, Kaysser T, Ursitti JA, Birkenmeier CS, Barker JE, and Bloch RJ. Na,K-ATPase in skeletal muscle: two populations of β-spectrin control localization in the sarcolemma but not partitioning between the sarcolemma and the transverse tubules. *J Cell Sci* 114: 751-762, 2001.

Wolitzky BA and Fambrough DM. Regulation of the (Na$^+$,K$^+$)-ATPase in cultured chick skeletal muscle. *J Biol Chem* 261: 9990-9999, 1986.

Zhang L, Parratt JR, Beastall GH, Pyne NJ, and Furman BL. Streptozotocin diabetes protects against arrhythmias in rat isolated hearts: role of hypothyroidism. *Eur J Pharmacol* 435: 269-276, 2002.

Zierler KL and Rabinowitz D. Effect of very small concentrations of insulin on forearm metabolism. Persistence of its action on potassium and free fatty acids without its effect on glucose. *J Clin Invest* 43: 950-962, 1964.